Oil and Gas Field
Development Techniques

WELL COMPLETION AND SERVICING

FROM THE SAME PUBLISHER

Basics of Reservoir Engineering
  R. COSSÉ

Drilling
  J.-P. NGUYEN

Drilling Data Handbook
  G. GABOLDE, J.-P. NGUYEN

Progressing Cavity Pumps
  H. CHOLET

Cement Evaluation Logging Handbook
  D. ROUILLAC

Cementing Technology and Procedures
  J. LECOURTIER, U. CARTALOS, Eds.

Log Data Acquisition and Quality Control
  PH. THEYS

Dictionnaire du forage et des puits
Dictionary of Drilling and Boreholes
anglais-français, français-anglais
  M. MOUREAU, G. BRACE

Well Testing: Interpretation Methods
  G. BOURDAROT

INSTITUT FRANÇAIS DU PÉTROLE PUBLICATIONS

Denis PERRIN
Head of Well Completion and Servicing Training
at ENSPM Formation Industrie
Professor at IFP School

in collaboration with
Michel CARON and **Georges GAILLOT**
ENSPM Formation Industrie

Oil and Gas Field
Development Techniques

WELL COMPLETION AND SERVICING

Translated from the French by Barbara Brown Balvet
and reviewed by Philippe Beun

1999

Translation (reviewed Edition) of
La production fond, D. Perrin
© 1995, Éditions Technip, Paris and
Institut français du pétrole, Rueil-Malmaison

© 1999, Editions Technip, Paris and
Institut français du pétrole, Rueil-Malmaison, France

All rights reserved. No part of this publication may be reproduced or transmitted in any form or by any means, electronic or mechanical, including photocopy, recording, or any information storage and retrieval system, without the prior written permission of the publisher.

ISBN 978-2-7108-0765-0
ISSN 1271-9048

The following authors took part in drafting this book :

Denis Perrin   Chapters 1, 2, 3 and 5
Michel Caron   Chapter 3
Georges Gaillot   Chapter 4

The autors would like to thank the instructors involved in the production seminars organized by ENSPM Formation Industrie as well as the lecturers in charge of courses at the IFP School. Their thanks are extended more particularly to those mentioned in the reference section specific to each chapter, whose input made this comprehensive study on production possible.

FOREWORD

Oil and gas field development and production include a host of specialty fields that can be divided up under four main headings:
- reservoir engineering
- drilling
- well completion and servicing
- surface production.

The four aspects are interdependent and oil and gas-related engineering, construction and operations entail the input of a great many operating and service company specialists, as well as equipment manufacturers and vendors. However, these professionals and technicians often lack a comprehensive grasp of all the specialized technology utilized in the process of producing oil and gas.

The aim of the four volumes presented under the title: "Oil and Gas Field Development Techniques" is to provide the basics of the technology and constraints involved in each of the four activities listed above in a condensed form, thereby allowing better interaction among the men and women who work in the different skill categories.

Furthermore, this introduction to oil and gas field technology is also designed to help specialists in other fields, e.g. information processing, law, economy, research, manufacturing, etc. It aims to enables them to situate their work in the context of the other skill categories and in this way make it easier for them to integrate their efforts in the professional fabric.

The four volumes partly recapitulate the contents of the training seminars organized by ENSPM-Formation Industrie to meet perceived needs, and also use some of the lecture material given to student engineers attending the IFP School.

A. Leblond
Former Head of Drilling, Production and Reservoir Engineering
at ENSPM-Formation Industrie

TABLE OF CONTENTS

Foreword VII

Chapter 1
INTRODUCTION TO COMPLETION

1.1 Main factors influencing completion design 1
    1.1.1 Parameters related to the well's purpose 2
        1.1.1.1 Exploration well (in the strict sense of the word) 2
        1.1.1.2 Confirmation or appraisal wells 3
        1.1.1.3 Development wells 3
    1.1.2 Parameters related to the environment 4
    1.1.3 Parameters related to drilling 4
        1.1.3.1 Type of drilling rig used 4
        1.1.3.2 Well profile 4
        1.1.3.3 Drilling and casing program 5
        1.1.3.4 Drilling in the pay zone(s) and drilling fluid 6
        1.1.3.5 Cementing the production casing 6
    1.1.4 Parameters related to the reservoir 6
        1.1.4.1 Reservoir pressure and its changes 6
        1.1.4.2 Interfaces between fluids and their changes 8
        1.1.4.3 Number of levels to be produced 9
        1.1.4.4 Rock characteristics and fluid type 9
        1.1.4.5 Production profile and number of wells required 9
    1.1.5 Parameters related to production 10
        1.1.5.1 Safety 10
        1.1.5.2 Flowing well or artificial lift 10
        1.1.5.3 Operating conditions 10
        1.1.5.4 Anticipated measurement, maintenance or workover operations 11
    1.1.6 Parameters related to completion techniques 11
    1.1.7 Synthesis: how completion is designed 11

1.2 Overall approach to a well's flow capacity 13
    1.2.1 Base equations 13
        1.2.1.1 The case of an oil flow 13
        1.2.1.2 The case of a gas flow 15
        1.2.1.3 Note 15
    1.2.2 Analysis of the different terms and resulting conclusions 16
        1.2.2.1 Decreasing the back pressure P_{BH} 17

	1.2.2.2 Slowing down the decline in P_R	20
	1.2.2.3 Increasing PI or C	20
1.2.3	Performance curves	21
1.2.4	Synthesis	22

1.3 Major types of completion configurations ... 23

- 1.3.1 **Basic requirements** ... 23
  - 1.3.1.1 Borehole wall stability ... 24
  - 1.3.1.2 Selectivity ... 24
  - 1.3.1.3 Minimizing restrictions in the flow path ... 24
  - 1.3.1.4 Well safety ... 24
  - 1.3.1.5 Flow adjustment ... 24
  - 1.3.1.6 Operations to be performed at a later date ... 24
- 1.3.2 **Pay zone-borehole connection: basic configurations** ... 25
  - 1.3.2.1 Open hole completions ... 25
  - 1.3.2.2 Cased hole completions ... 26
- 1.3.3 **Main configurations of production string(s)** ... 27
  - 1.3.3.1 Conventional completions ... 27
  - 1.3.3.2 Tubingless completions ... 30
  - 1.3.3.3 Miniaturized completions ... 31

1.4 Main phases in completion ... 32

- 1.4.1 **Checking and conditioning the borehole** ... 33
- 1.4.2 **Remedial cementing** ... 33
- 1.4.3 **Re-establishing pay zone-borehole communication** ... 33
- 1.4.4 **Well testing** ... 34
- 1.4.5 **Treating the pay zone** ... 34
- 1.4.6 **Equipment installation** ... 34
- 1.4.7 **Putting the well on stream and assessing performance** ... 34
- 1.4.8 **Moving the rig** ... 35
- 1.4.9 **Operations to be performed at a later date: measurements, maintenance, workover and abandonment** ... 35

Chapter 2
CONNECTING THE PAY ZONE AND THE BOREHOLE

2.1 Drilling and casing the pay zone ... 37

- 2.1.1 **Well safety** ... 37
- 2.1.2 **Fluids used to drill in the pay zone** ... 38
  - 2.1.2.1 Constraints ... 38
  - 2.1.2.2 Completion fluids ... 40
- 2.1.3 **Drilling and casing diameters** ... 43
- 2.1.4 **Casing and cementing** ... 43

2.2 Evaluating and restoring the cement job 44
2.2.1 Evaluating the cement job 44
2.2.1.1 Major flaws encountered after primary cementing 44
2.2.1.2 Principal methods of evaluating a cement job 45
2.2.2 Remedial cementing 47
2.2.2.1 Introduction 47
2.2.2.2 Squeeze techniques 49
2.2.2.3 Squeeze procedures and corresponding tool strings 51
2.2.2.4 Implementation (low-pressure squeeze) 56

2.3 Perforating 59
2.3.1 Shaped charges 59
2.3.1.1 Principle 59
2.3.1.2 The API RP 43 standard 60
2.3.2 Main parameters affecting the productivity of the zone produced by perforating 63
2.3.2.1 Number of effective perforations 63
2.3.2.2 Distribution of perforations over the producing zone (partial penetration effect) 63
2.3.2.3 Perforation penetration 63
2.3.2.4 Characteristics of the crushed zone 64
2.3.2.5 Number of shot directions 64
2.3.2.6 Perforation diameter 65
2.3.3 Perforating methods and corresponding types of guns 65
2.3.3.1 Overbalanced pressure perforating before equipment installation 66
2.3.3.2 Underbalanced pressure perforating after equipment installation 68
2.3.3.3 TCP perforating (Tubing Conveyed Perforator) 70
2.3.4 Specific points in the operating technique 72
2.3.4.1 Safety 72
2.3.4.2 Perforation depth adjustment 73
2.3.4.3 Cleaning the perforations 74
2.3.4.4 Monitoring the results 77

2.4 Treating the pay zone 77
2.4.1 Problems encountered 77
2.4.1.1 Phenomena pertaining to insufficient consolidation 77
2.4.1.2 Phenomena pertaining to insufficient productivity 78
2.4.1.3 Main causes of problems 79
2.4.1.4 Means of diagnosis 80
2.4.2 Main types of remedial action for poor consolidation: sand control 81
2.4.2.1 Why control sand? 81
2.4.2.2 Sand control processes 82
2.4.2.3 Gravel packing in a perforated cased hole 87
2.4.3 Main types of remedial action for insufficient productivity: well stimulation 88
2.4.3.1 Principal types of well stimulation 88
2.4.3.2 Acidizing 91
2.4.3.3 Hydraulic fracturing 95

2.5 The special case of horizontal wells 107
2.5.1 Advantages in producing reservoirs 107
2.5.1.1 Low permeability formation 107

TABLE OF CONTENTS

    2.5.1.2 Thin formation 108
    2.5.1.3 Plugged formation 108
    2.5.1.4 Effect of turbulence 108
    2.5.1.5 Critical flow rate (in relationship with coning) 109
    2.5.1.6 Insufficiently consolidated formation 109
    2.5.1.7 Naturally fractured, heterogeneous formation, etc. 109
    2.5.1.8 Secondary recovery 110
  2.5.2 **Problems specific to the pay zone-borehole connection** 110
    2.5.2.1 Pay zone-borehole connection configuration 110
    2.5.2.2 Running in and cementing the liner 111
    2.5.2.3 Perforating 111
    2.5.2.4 Sand control 112
    2.5.2.5 Stimulation 113
    2.5.2.6 Configuration of production string(s) 113

Chapter 3
THE EQUIPMENT OF NATURALLY FLOWING WELLS

3.1 General configuration of flowing well equipment 115

3.2 The production wellhead 117
  3.2.1 **Hanging (and securing) the tubing** 118
  3.2.2 **The Christmas tree (Xmas tree)** 118

3.3 The production string or tubing 122
  3.3.1 **Tubing characteristics** 122
    3.3.1.1 Nominal diameter and geometrical characteristics 122
    3.3.1.2 Nominal weight 123
    3.3.1.3 Grades of steel and metallurgical characteristics 124
    3.3.1.4 Connections, threads 125
    3.3.1.5 Mechanical characteristics of tubing pipe 126
  3.3.2 **Choosing the tubing** 128
    3.3.2.1 Determining the nominal pipe diameter 128
    3.3.2.2 Determining the grade and the nominal weight 130
    3.3.2.3 Determining the connection and the metallurgy 130

3.4 Packers 131
  3.4.1 **Packer fluids (or annular fluids)** 131
  3.4.2 **The main packer types** 132
  3.4.3 **Choosing the packer** 132
  3.4.4 **Permanent production packers** 133
    3.4.4.1 Description of the 415 D packer 133
    3.4.4.2 Setting the 415 D packer on the electric cable 133

3.4.4.3 Setting the 415 D packer on a pipe string	135
3.4.4.4 Connection between the tubing and the 415 D packer	135
3.4.4.5 Drilling out the 415 D packer	138
3.4.4.6 Advantages and drawbacks of permanent packers	138
3.4.5 Retrievable packers	138
3.4.5.1 Hydraulically set retrievable packers	139
3.4.5.2 Mechanically set retrievable packers	142

3.5 Downhole equipment ... 142

3.5.1 Circulating devices	142
3.5.1.1 Sliding sleeve circulating valve	143
3.5.1.2 Side pocket mandrel	143
3.5.1.3 Ported landing nipple	145
3.5.1.4 Conclusion, comparison	145
3.5.2 Landing nipples	145
3.5.2.1 Full bore simple landing nipples	146
3.5.2.2 Full bore selective landing nipples	148
3.5.2.3 Full bore top no-go landing nipples	149
3.5.2.4 Bottom no-go landing nipples	150
3.5.2.5 Landing nipple summary table	150
3.5.2.6 Using several landing nipples in the same tubing	152
3.5.3 Other downhole equipment	153

3.6 Subsurface safety valves ... 153

3.6.1 Subsurface Controlled Subsurface Safety Valves (SSCSV)	154
3.6.1.1 Pressure differential safety valves	154
3.6.1.2 Pressure operated safety valves	155
3.6.2 Surface Controlled Subsurface Safety Valves (SCSSV)	156
3.6.2.1 Wireline Retrievable safety valves (WLR)	157
3.6.2.2 Tubing Retrievable safety valves (TR)	157
3.6.2.3 Combination safety valves	159
3.6.2.4 Subsurface tubing-annulus safety valves	159
3.6.2.5 Other safety valves	159

3.7 Running procedure ... 161

3.7.1 Preliminary operations	161
3.7.1.1 Reconditioning the wellhead	161
3.7.1.2 Checking the borehole	163
3.7.1.3 Cased hole logging	163
3.7.1.4 Reconditioning the BOPs	164
3.7.1.5 Re-establishing the pay zone-borehole connection : perforating	164
3.7.2 Running subsurface equipment in a well equipped with a permanent packer (packer set prior to running the tubing)	164
3.7.2.1 Setting the packer (and the below-packer extension)	164
3.7.2.2 Assembling and running the equipment (and testing while running in hole)	165
3.7.2.3 Inserting the subsurface safety valve landing nipple and continuing to run in	166
3.7.2.4 Spacing out the production string and completing tubing equipment	166
3.7.2.5 Landing the tubing hanger in the tubing head	168
3.7.2.6 Testing the production string and the annulus	168

- 3.7.3 **Main differences in running subsurface equipment when there is a hydraulic packer (run in directly on the tubing)** 169
    - 3.7.3.1 Assembling and running the equipment (and testing while running in hole) 169
    - 3.7.3.2 Inserting the SCSSV landing nipple and finishing to run the equipment 169
    - 3.7.3.3 Partial testing of the production string 169
    - 3.7.3.4 Setting the hydraulic packer, landing the tubing hanger and testing 170
- 3.7.4 **Installing the Christmas tree and bringing the well on stream** 170
    - 3.7.4.1 Replacing the BOPs by the Christmas tree 170
    - 3.7.4.2 Testing the production wellhead 171
    - 3.7.4.3 Pumping in annular and clearing fluids 172
    - 3.7.4.4 Setting and testing the SCSSV 172
    - 3.7.4.5 Clearing the well 173
- 3.7.5 **Final completion report** 174

Chapter 4
ARTIFICIAL LIFT

- **4.1 Pumping** 175
    - 4.1.1 **Principle and types of pumping** 175
    - 4.1.2 **Sucker rod pumping** 176
        - 4.1.2.1 Principle 176
        - 4.1.2.2 Downhole equipment 178
        - 4.1.2.3 Choosing pumping parameters 186
        - 4.1.2.4 Pumping Units (PU) 188
    - 4.1.3 **Submerged centrifugal pumping** 195
        - 4.1.3.1 Component parts of the centrifugal pump 195
        - 4.1.3.2 The surface control equipment 197
        - 4.1.3.3 Selecting a pump 198
    - 4.1.4 **Hydraulic pumping** 199
        - 4.1.4.1 General introduction 199
        - 4.1.4.2 Principle of the plunger pump 199
        - 4.1.4.3 Well equipment 201
        - 4.1.4.4 Principle of the jet pump 204
        - 4.1.4.5 The turbine pump 205
    - 4.1.5 **The Moyno pump** 206
    - 4.1.6 **Measurements on pumped wells** 207
        - 4.1.6.1 Measuring levels in the annulus (echometer) 209
        - 4.1.6.2 Dynamometric measurements at the polished rod 209
        - 4.1.6.3 Measuring intensity in submerged centrifugal pumping 211
    - 4.1.7 **Defining a pumping installation** 215
        - 4.1.7.1 A sucker rod pumping system 215
        - 4.1.7.2 Submerged centrifugal pumping 219
        - 4.1.7.3 Hydraulic plunger pumping 221
        - 4.1.7.4 Hydraulic jet pumping 221

4.2 Gas lift 222

4.2.1 Principle and types of gas lift 222
- 4.2.1.1 Principle 222
- 4.2.1.2 Types of gas lift 223

4.2.2 Regarding the well 225
- 4.2.2.1 Operating conditions in continuous gas lift 225
- 4.2.2.2 Pressure gradients in producing wells 227
- 4.2.2.3 Unloading a well at start up with unloading valves 230
- 4.2.2.4 Gas-lift valve technology 233
- 4.2.2.5 Tubing equipment specific to gas lift 236
- 4.2.2.6 Kickover or positioning tools 240

4.2.3 Surface equipment for a gas lifted well 241
- 4.2.3.1 Injection system 241
- 4.2.3.2 Measurements 247

4.3 Choosing an artificial lift process 247

4.3.1 Economic criteria 247
4.3.2 Technical criteria 248
4.3.3 Making a decision 248
4.3.4 Main advantages and drawbacks of artificial lift processes 249
- 4.3.4.1 Sucker rod pumping 249
- 4.3.4.2 Submerged centrifugal pumping 250
- 4.3.4.3 Hydraulic pumping 250
- 4.3.4.4 Continuous gas lift 251

Chapter 5
WELL SERVICING AND WORKOVER

5.1 Main types of operations 253

5.1.1 Measurement operations 254
- 5.1.1.1 At the wellhead 254
- 5.1.1.2 In the tubing 255
- 5.1.1.3 At the bottomhole 255

5.1.2 Maintenance operations 255
- 5.1.2.1 On the wellhead 255
- 5.1.2.2 In the tubing and its equipment 255
- 5.1.2.3 At the bottomhole and on the pay zone-borehole connection ... 256

5.1.3 Workover operations 256
- 5.1.3.1 Equipment failure 256
- 5.1.3.2 Modifications in production conditions 257
- 5.1.3.3 Restoration or modification of the pay zone-borehole connection ... 257
- 5.1.3.4 Change in the purpose of the well 257
- 5.1.3.5 Fishing 258

5.2 Light operations on live wells ... 258
5.2.1 Wireline work ... 258
5.2.1.1 Principle and area of application ... 258
5.2.1.2 Surface equipment ... 259
5.2.1.3 The wireline tool string ... 265
5.2.1.4 Wireline tools ... 269
5.2.2 Pumping ... 276

5.3 Heavy operations on live wells ... 276
5.3.1 Coiled tubing ... 276
5.3.1.1 Principle and area of application ... 276
5.3.1.2 Coiled tubing equipment ... 278
5.3.1.3 Operating considerations ... 282
5.3.2 Snubbing ... 283
5.3.2.1 Principle and area of application ... 283
5.3.2.2 Snubbing equipment ... 283
5.3.2.3 Operating considerations ... 292

5.4 Operations on killed wells ... 293
5.4.1 Means of acting on killed wells ... 293
5.4.2 General procedure of an operation ... 294
5.4.2.1 Preparing the well (before the servicing or workover unit arrives) ... 294
5.4.2.2 Putting the well under provisional safe conditions (before rigging up the servicing or workover unit) ... 294
5.4.2.3 Installing the servicing or workover unit ... 295
5.4.2.4 Killing the well ... 295
5.4.2.5 Replacing the Christmas tree with the BOPs ... 295
5.4.2.6 Removing completion equipment ... 295
5.4.2.7 Downhole operations, recompletion, replacing BOPs by the Christmas tree and start up .. 296
5.4.2.8 Moving out the servicing or workover unit ... 296
5.4.3 Considerations on killing the well ... 296
5.4.3.1 Killing by circulating ... 296
5.4.3.2 Killing by squeezing ... 297
5.4.3.3 Observing the well ... 298
5.4.3.4 Final killing phase ... 298
5.4.4 Depleted reservoirs ... 298
5.4.4.1 Losses and formation damage ... 299
5.4.4.2 Start up problems ... 299
5.4.5 Fishing tools ... 299

5.5 Special cases ... 303
5.5.1 Operations on horizontal wells ... 303
5.5.2 Operations on subsea wells ... 305

Bibliography ... 309
Index ... 313

Chapter 1

INTRODUCTION TO COMPLETION

The word "completion" itself means conclusion, and more particularly in the case we are concerned with, the conclusion of a borehole that has just been drilled. Completion is therefore the link between drilling the borehole as such and the production phase.

As a result, completion involves all of the operations designed to make the well produce, in particular connecting the borehole and the pay zone, treating the pay zone (if any treatment), equipping the well, putting it on stream and assessing it. By pay zone we mean a zone consisting of reservoir rocks which contain oil and/or gas that can be recovered.

Generally speaking, we usually consider that certain measurement and maintenance operations in the well along with any workover jobs that might be required also come under the heading of completion.

Completion is highly dependent on the phases that precede and follow it and is often even an integral part of them. We can therefore say that completion begins with well positioning and ends only at well abandonment.

Whatever the operational entity in charge of well completion and workover, its actions are greatly influenced by the way the well has been designed and drilled and by the production problems the reservoir might cause. The "completion man" will therefore have to work in close cooperation with the "driller" (who may both work in one and the same department), and also with reservoir engineers and production technical staff.

1.1 MAIN FACTORS INFLUENCING COMPLETION DESIGN

There are many factors influencing completion, so we will examine the main ones here broken down into six categories, before making a synthesis. The six types of parameters are related to:

- the well's purpose
- the environment
- drilling
- the reservoir

- production
- completion techniques.

1.1.1 Parameters related to the well's purpose

The purpose of drilling can vary depending on the well, with a distinction basically made between:
- the exploration well (in the strict sense of the word, i.e. the first well)
- confirmation or appraisal wells
- development wells.

1.1.1.1 Exploration well (in the strict sense of the word)

The prime objective of this well is to define the nature of the fluids (water, oil or gas) in the reservoir rock and to get preliminary data on the reservoir, i.e. to make measurements.

As a priority, we try to define:
- the nature and characteristics of the fluids in place in the reservoir (including the water)
- the characteristics of the pay zone and more particularly the initial pressure, the temperature and the approximate permeability and productivity.

Generally, this involves:
- making a number of measurements by means of tools lowered into the well on an electric cable (wireline logging)
- running in a temporary test string in order to carry out production testing.

During this production testing, or well testing, the aim is mainly to take samples and record the variation in bottomhole pressure following a variation in flow rate (the well is produced at a settled production rate Q or shut in after a settled production rate Q).

Subsequently, the well may be suspended or abandoned.

Let us point out that complications arise in working out the testing and abandonment program for this exploration well because the required data are generally very incompletely known and often available only at the last minute.

The information obtained from this exploration well will help to complete the data already available from geology, geophysics, and so forth.

Based on this a decision must be made:
- either not to develop the reservoir, or
- to develop the reservoir, or
- to drill one or more further wells to obtain additional information (see following section).

1.1.1.2 Confirmation or appraisal wells

The purpose of these wells is to refine or complete the data from the exploration wells. Another aim, provided time is not too limited by safety considerations, is to determine the off-wellbore reservoir characteristics, particularly:
- the off-wellbore permeability
- the existence of heterogeneity, discontinuity or faults
- the reservoir boundaries, a possible water drive.

In order to do this, well testing is performed, usually over a longer period of time than for an exploration well.

All the data obtained on these different wells are collected and used to make the first correlations between wells, thereby giving a picture on the scale of the field rather than just the well.

The following steps are then necessary:
- Work out development schemes with corresponding production forecasts.
- Make the decision whether to develop the field, and if the answer is yes, choose among the schemes to draw up the development project accordingly.

1.1.1.3 Development wells

The main purpose of these wells is no more to make measurements but to bring the field on stream, with priority going to their flow capacity. However, it is also important to test this type of well to:
- Assess the condition of the well and check how effective the completion has been.
- If need be, obtain further information about the reservoir.

There are different types of development wells: production, injection and observation wells.

A. Production wells

These are the most numerous, their aim is to optimize the productivity-to-price ratio.

B. Injection wells

These are much less numerous, but are often crucial in producing the reservoir. In particular, some injection wells are used to maintain reservoir pressure and others to get rid of an unwanted fluid.

C. Observation wells

There are usually few or none of this type on a field. They are completed to monitor variations in reservoir parameters (e.g. interface between fluids, pressure, etc.). Usually, observation wells are wells initially drilled for production or injection, then found unsuitable for that purpose.

Over a period of time, the same well can be used in different ways, for example production, then injection (possibly after a certain shut-in period).

This book deals mainly with the subject of production wells.

1.1.2 Parameters related to the environment

There may be constraints on operations due to the country or site where the well is located, whether on land (plain or mountain, desert, agricultural or inhabited area, etc.) or offshore (floating platform, development from a fixed platform or by subsea wellhead). Restrictions may involve:

- difficulties in obtaining supplies
- available space
- available utilities
- safety rules that have to be enforced
- certain operations that may or may not be possible.

Meteorological and, if relevant, oceanographic conditions must also be taken into account.

1.1.3 Parameters related to drilling

1.1.3.1 Type of drilling rig used

Although wells are sometimes completed with a specific rig that replaces the drilling rig, the same rig is often used for drilling and for completion. The following points must be taken into consideration:

- the characteristics of the rig
- the type of equipment provided on it
- any additional units (e.g. cementing) that may be available on it.

In fact, it is better to choose the drilling rig from the outset with due consideration given to the requirements specific to completion.

1.1.3.2 Well profile

Well deviation is linked to a cluster of wellheads (on land, on a platform offshore or on the seabed), and/or to reservoir engineering considerations (recovery efficiency, horizontal drilling, etc.). Deviation may limit or even rule out the choice of some equipment or technique used to work in the well.

1.1.3.3 Drilling and casing program (Fig. 1.1)

For a development well, the most important thing is to have a borehole with a big enough diameter to accommodate the equipment that will be installed in it. In contrast, when the pay zone drilling diameter is increased above and beyond what is required for the production equipment, it does not boost the well's flow capacity very much.

What we are concerned with here is therefore the inside diameter that is effectively usable in the well once all the drilling and casing phases have been concluded.

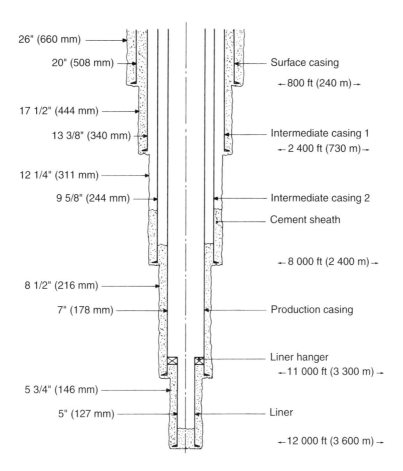

Fig. 1.1 Diameters available according to the drilling and casing program (*Source:* ENSPM Formation Industrie).

Since the diameter depends on the initial drilling program, this explains the saying that is sometimes used: "Completion begins with the first turn of the bit".

As a result, the drilling and casing program must be optimized taking both drilling and production requirements into account, without losing sight of the flow capacity versus investment criterion.

1.1.3.4 Drilling in the pay zone(s) and drilling fluid

From the time the drilling bit reaches the top of the reservoir and during all later operations, reservoir conditions are disturbed. Because of this, problems may arise in putting the well on stream.

In particular, the pay zone may be damaged by the fluids used in the well (drilling fluid, cement slurry, etc.), and this means reduced productivity.

Depending on the case, productivity can be restored relatively easily (generally true for carbonate formations: limestone, dolomites, etc.). It may prove to be difficult or even impossible for sandstone formations. In any case, it requires costly treatment in terms of rig time and of the treatment itself.

Formation damage should therefore not be seen simply in terms of the cure but also in terms of prevention, especially when formation plugging is very expensive or impossible to solve. As a result, the choice of fluid used to drill the pay zone is critical.

While drilling in a number of other factors can influence production start up. Let us mention problems of wellbore stability and accidental fracturing (for example during a kick control) which may cause unwanted fluids to be produced.

Generally speaking, it is essential to find out exactly what was done during this phase. Due consideration should more particularly be given to the events (fluid inflows, losses, etc.) that may have occurred and how they were dealt with.

1.1.3.5 Cementing the production casing

A good seal provided by the sheath of cement between the formation and the production casing is a very important parameter, mainly because of reservoir performance (see following section). It is therefore necessary to examine the way this cementing job is carried out and tested.

1.1.4 Parameters related to the reservoir

1.1.4.1 Reservoir pressure and its changes

What is of interest here is not so much the initial reservoir pressure as how it varies over time. The fact is that reservoir pressure is a key parameter in the well's natural flow capability. If the reservoir pressure is or becomes insufficient to offset production pressure drawdown

1. INTRODUCTION TO COMPLETION

(particularly the hydrostatic pressure of the fluid column in the well and pressure losses), it is then necessary to install a suitable artificial lift system such as pumping the fluids or lightening them by gas injection in the lower part of the tubing (gas lift).

If a reasonably accurate estimate of future requirements in this area can be made at the time of initial completion, an attempt is made to take them into consideration when completion equipment is chosen. Such a procedure can make later workover easier or unnecessary.

The change in reservoir pressure is physically related to cumulative production (rather than directly to time) and to the drive mechanism(s) involved (Fig. 1.2). To go from cumulative production to time, the production rate is used, which is in turn related to the number of producing wells and the flow rate per well. Let us point out that there may be local regulations restricting the production rate or the flow rate per well.

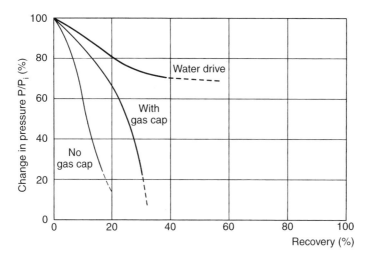

Fig. 1.2 Change in reservoir pressure versus cumulative production (*Source:* R. Cossé, *Reservoir Engineering,* Éditions Technip, Paris, 1989).

Injection wells may supplement the action of natural drive mechanisms such as one-phase expansion, solution gas drive, gas cap drive or water drive. The injected fluid maintains pressure (or slows down the decline of the reservoir pressure) and in addition flushes out the oil (sweeping effect). Although the two functions can not be dissociated in practice, one of them (maintaining pressure or sweeping effect) can more particularly justify this type of well. Mostly water is injected, but gas may sometimes also have to be injected.

Do not confuse injection wells where water (or gas) is injected into the reservoir to enhance drive (assisted drive) with:

- Injection wells (in whatever layer) designed to get rid of unwanted fluids on the surface (oily water, etc.), consequently called "dump wells".

- Wells where gas is directly injected into the tubing (through the annulus) to help lift the oil up to the surface (production assisted by gas lift, see Chapter 4).

1.1.4.2 Interfaces between fluids and their changes

The existence of interfaces between fluids, in particular when they are not controlled, causes a decrease in target fluid productivity at the same time as an increase in unwanted fluids (water and gas for an oil reservoir, water for a gas reservoir).

Additionally, since these unwanted fluids get into the well, they must be brought up to the surface before they can be disposed of. They therefore not only penalize well productivity but are also instrumental in decreasing reservoir pressure.

This interface problem is more particularly critical when the viscosity of the target fluid is more or less the same (light oil and water) or even much greater (heavy oil and water, oil and gas) than that of the unwanted fluid.

The interfaces vary with time, for example locally around a well, by a suction phenom-enon causing a cone (coning) which is related to the withdrawal rate (Fig. 1.3). They can also vary throughout the reservoir depending on the amount of fluid that has already been withdrawn, allowing a gas cap or an aquifer, to expand or the pressure to pass below the bubble point.

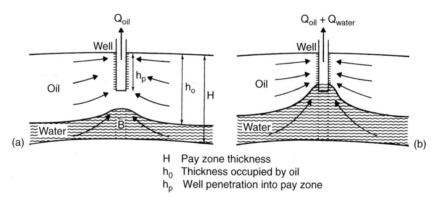

H   Pay zone thickness
h_o Thickness occupied by oil
h_p Well penetration into pay zone

Fig. 1.3 Coning (*Source:* After R. Cossé, *Reservoir Engineering*, Éditions Technip, Paris, 1989).
a. Stable cone. b. Water encroachment.

This causes a variation in the percentage of oil, gas and water in the produced effluent.

The reservoir configuration can accentuate the phenomenon. For example, let us mention the case of relatively thin reservoirs where it is not possible to locate the pay zone to borehole connection far enough from the interface. Another instance is when there is a highly permeable drain (fissure, etc.) which makes the influx of unwanted fluid easier without letting the target fluid be drained out of the matrix.

1.1.4.3 Number of levels to be produced

When there are several formations to be developed (by production or injection), it may be desirable to drill a well that can produce several of them rather than just one, whether commingled production will be allowed or not.

A suitable completion program can be examined based on what is permitted by local regulations and without disregarding drilling or reservoir considerations. However, even if the initial investment is often lower, there are no miracles from a technical standpoint. Careful attention must be paid to extra risks that are taken. In particular, workover jobs may be:
- More frequent because equipment is more complex.
- More difficult because formations are often depleted at different rates.

1.1.4.4 Rock characteristics and fluid type

Rock characteristics and the type of reservoir fluids will directly influence completion, especially with respect to the well's flow capacity, the type of formation treatments that have to be considered and the production problems that have to be dealt with. Let us mention the following parameters in particular:
- the nature and composition of the rocks
- the degree of reservoir consolidation
- the extent of reservoir damage
- the temperature
- the fluid's viscosity
- the fluid's corrosive or toxic properties
- the fluid's tendency to emulsify or lay down deposits.

1.1.4.5 Production profile and number of wells required

Choosing the production profile and therefore determining the number of wells required are the result of a number of factors. The decision is mainly based on the following points:
- the size of the reservoir, the existing permeability barriers, the well drainage radius
- the drive mechanisms
- the individual flow capacity which is related to the characteristics of the reservoir and of the oil and/or gas, to fluid interface problems, and to the drive mechanisms and artificial lift method involved
- any local regulations concerning depletion rate, maximum flow rate per well or multilayer reservoir production
- siting constraints on the surface
- economic factors such as development costs, operating expenses, oil and/or gas sales price, taxation, etc.

1.1.5 Parameters related to production

1.1.5.1 Safety

Well equipment must comply with:
- national regulations
- local regulations
- company rules and regulations.

Let us simply mention the following points, since each one will be dealt with later on:
- the number and position of surface safety valves
- whether there is a tubing, a packer (a downhole seal between the casing and the tubing) or, in case of a tubingless completion, a kill string (designed to "kill" the well, i.e. allowing a control fluid such as brine or drilling fluid to be pumped in to exert a hydrostatic pressure that is greater than the reservoir pressure)
- whether there are subsurface safety valves or devices (tubing or tubing + annulus) and what type they are
- the type and specific gravity of the annular fluid.

1.1.5.2 Flowing well or artificial lift

The advisability of artificial lift in the future must be examined right from the beginning. If the answer is affirmative, an attempt should be made to determine which process(es) might effectively be used. The aim is to reserve the required space both in the well and on the surface (especially important for offshore development). If gas lift is the artificial lift method choosen, it is usually possible to pre-equip the well accordingly so that workover will not be necessary.

Basically, a well's flow capability depends on the natural change in bottomhole pressure versus cumulative production. However, it also depends on:
- the required flow rate
- whether there is some sort of artificial drive (e.g. injection well)
- the change over time in the composition of the effluent that flows into the well (gas or water influx, for example)
- the characteristics of the pay zone, particularly in the vicinity of the wellbore (pressure losses in the formation)
- the diameter of the tubing and flowlines (pressure losses)
- the pressure required for surface processing (separation pressure).

1.1.5.3 Operating conditions

In addition to the separation pressure that was mentioned above, the following points also have to be taken into consideration:

- the immediate environment, such as the available space and utilities (especially in choosing an artificial lift method)
- operating problems, mainly those related to the type of fluid (e.g. erosion, corrosion, deposits), or to the temperature (true for deep wells, steam injection wells, etc.)
- operations that are due to be carried out on or in the well (see following section).

1.1.5.4 Anticipated measurement, maintenance or workover operations

During a well's lifetime, measurements must be made to be sure that the well is producing properly and to monitor the way the reservoir is behaving.

Maintenance and repair operations also must be made possible for the completion system that has been chosen and installed.

Finally, some or all of the completion system may have to be modified to match changes in certain parameters (see earlier: reservoir pressure and therefore flow capability, position of interfaces, etc.). Changes can also be made in the use that the well is put to (change in the produced formation, turning a producer into an injector, etc.).

The type of completion chosen must allow these operations to be carried out directly or to perform workover as needed with risks and costs reduced to a strict minimum.

1.1.6 Parameters related to completion techniques

Based on the parameters mentioned above, a number of choices have to be made for completion, for example:

- the general configuration
- the pay zone to borehole connection
- treatment of the formation if required
- the different pieces of well equipment installed
- possible artificial lift
- the operating procedure used to carry out completion
- procedures for operations to be performed on the well at a later date.

However, on the one hand these choices are interdependent, and on the other a choice that is made in accordance with one parameter (in drilling for example) may not be compatible with another parameter (in reservoir engineering for example). As a result, the completion system that is chosen is the result of a trade-off.

1.1.7 Synthesis: how completion is designed

The main purposes of a well are generally decided by the Company's operations management and the reservoir engineering department:

- For exploration and appraisal wells this mainly involves the level(s) that are to be tested, and the type and duration of the tests that are to be run.
- For development wells this basically involves the level(s) that are to be produced and the production or injection profile required for the wells.

Based on the above, particularly for development wells, the problem is to design the best possible completion in order to:

- optimize productivity or injectivity performance during the well's complete lifetime
- make sure that the field is produced reliably and safely
- optimize the implementation of an artificial lift process
- optimize equipment lifetime
- make it possible to change some or all of the well's equipment at a later date without too much difficulty so that it can be adapted to future operating conditions
- minimize initial investment, operating costs and the cost of any workover jobs.

This may mean a compromise in the drilling and casing program or in operating conditions or even that the objectives have to be modified if they prove to be unattainable.

As we saw previously, the data required to set up a completion system are very numerous. We will only list the most important constraints and parameters below:

- local constraints (regulations, environment, etc.)
- the type of effluents and their characteristics
- the reservoir and its petrophysical characteristics
- the number of producing formations, each one's expected productivity and the interfaces
- the available diameter and the borehole profile
- whether it is necessary to proceed to additional operations (well stimulation, sand control, etc.)
- whether it is necessary to implement techniques to maintain reservoir pressure (water, gas, solvent or miscible product injection) or to lift the effluents artificially (gas lift, pumping, nitrogen or carbon dioxide injection) immediately or at a later date
- the eventuality of having to do any work on the pressurized well during the production life by wireline, or with a concentric tubular (coiled tubing or snubbing).

Completion design is based on this body of data, so every effort must be made to be sure no important point has been disregarded, since incomplete or wrong data might lead to poor design.

The job is not an easy one since:

- These data are very numerous and may be interrelated.
- Some of them are not very accurately known when completion is designed (sometimes not even when completion is being carried out).
- Some of them are contradictory.
- Some of them are mandatory, while others can be subject to compromise.

1.2 OVERALL APPROACH TO A WELL'S FLOW CAPACITY

A well's flow capacity is a crucial parameter because of its economic consequences. It is therefore important to endeavor to assess it. However, it should be pointed out that the flow capacity evolves over time and unfortunately tends to decrease.

1.2.1 Base equations

A well's flow rate depends on:
- the existing pressure difference, i.e. the reservoir pressure (P_R) and the back pressure exerted downhole (P_{BH}: bottomhole pressure)
- parameters that involve the type of reservoir and in-place fluids.

1.2.1.1 The case of an oil flow

Here, provided that there is no free gas, that the flow can be considered to be of the radial cylindrical steady-state type and that the fluid velocity is not too great in the vicinity of the wellbore, the flow equation can be simplified to:

$$Q = PI(P_R - P_{BH})$$

where the productivity index (PI) mainly depends on the viscosity of the fluid, the permeability of the formation itself, the disturbances in the vicinity of the wellbore and the thickness of the reservoir.

In fact the actual productivity index (PI) can be compared with the theoretical productivity index (PI_{th}) of a vertical well at the level of the formation that would have been drilled under ideal conditions. By this we mean without having interfered with the reservoir characteristics in the vicinity of the wellbore (permeability especially) and with no restrictions on the connection between the reservoir and the wellbore.

Here the theoretical productivity index is as follows (Diagram 1.1):

$$PI_{th} = \alpha \frac{hk}{\mu \ln \frac{R}{r_w}}$$

with:
- α numerical coefficient depending, among other things, on the units that are used
- h reservoir thickness
- k reservoir permeability
- μ viscosity of the fluid in the reservoir
- R well drainage radius
- r_w wellbore radius

1. INTRODUCTION TO COMPLETION

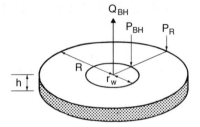

Diagram 1.1 (*Source:* ENSPM Formation Industrie).

As far as the real well is concerned, all of the disturbances in the vicinity of the wellbore (skin effect) are lumped together under the term "S" (skin factor) in the following way:

$$PI = \alpha \frac{hk}{\mu \left(\ln \frac{R}{r_w} + S \right)}$$

Furthermore, the flow efficiency (*Fe*) is defined as the ratio between the actual flow rate and the theoretical flow rate that the "ideal" well would have under the same bottomhole pressure conditions:

$$Fe = \frac{Q}{Q_{th}} = \frac{PI}{PI_{th}} = \frac{\ln \frac{R}{r_w}}{\left(\ln \frac{R}{r_w} + S \right)}$$

In practice $\ln R/r_w$ often ranges between 7 and 8, hence the simplified form (under the conditions laid down at the beginning of this section):

$$Fe = \frac{PI}{PI_{th}} \approx \frac{7}{7+S} \quad \text{to} \quad \frac{8}{8+S}$$

A skin factor of 7 to 8 therefore corresponds to a flow capacity that has been divided by two. A factor of 14 to 16 therefore means it has been reduced by two-thirds. In contrast, a skin factor of −3.5 to −4 (after well stimulation for example) means that it has been doubled.

The skin factor, often considered as the effect of plugging in the vicinity of the wellbore, is in fact the result of several factors:

- S_{fp} due to formation plugging per se.
- S_p due to the perforations themselves (considering only the linear law part of the curve of $P_r - P_{BH}$ as a function of Q).
- S_t due to the effect of turbulence in the perforations or in the immediate vicinity of the wellbore (deviation with respect to the linear law). Attention should be paid to the fact that, contrary to the other terms which are independent of the flow rate, S_t varies with it (Fig. 1.4).

S_{pp} due to the partial penetration effect when the formation has not been perforated throughout its thickness.

S_d due to the deviation effect, which is generally negligible unless the well is considerably deviated or horizontal; it should be noted that S_d is (zero or) negative and therefore enhances flow.

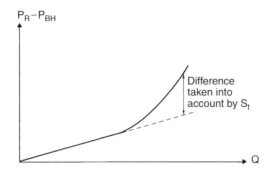

Fig. 1.4 Turbulence effect (*Source:* ENSPM Formation Industrie).

A clear distinction must be made between the different factors, since the different "cures" will require different techniques.

1.2.1.2 The case of a gas flow

Here, setting up an equation is more complex. The following empirical equation is often used (especially by Americans):

$$Q_{std} = C(P_R^2 - P_{BH}^2) \quad \text{with} \quad 0.5 < n < 1$$

In this formula Q_{std} is a volume flow rate under what are termed standard conditions and C is mainly dependent, as is PI, on the viscosity of the fluid, the mean permeability and reservoir thickness.

Actually, provided that the flow can be considered to be of the radial cylindrical steady-state type for a gas well, the following equation (termed a quadratic equation) can be established:

$$P_R^2 - P_{BH}^2 = AQ_{std} + BQ_{std}^2$$

Here A is once again mainly dependent on the viscosity of the fluid, the mean permeability and reservoir thickness, and the higher the flow velocity, the less negligible the term BQ^2 is.

1.2.1.3 Note

The formulas for cases of transient or of multiphase flow are not dealt with here since they are much more complex to express. Even so, the factors mentioned above remain valid.

1. INTRODUCTION TO COMPLETION

1.2.2 Analysis of the different terms and resulting conclusions

For production to occur, the bottomhole pressure must be lower than the formation pressure. At the same time bottomhole pressure is no more than the back pressure caused by the whole downstream circuit (Fig. 1.5).

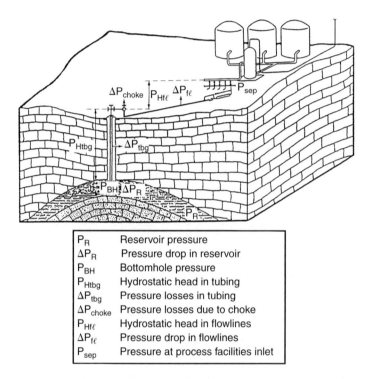

P_R Reservoir pressure
ΔP_R Pressure drop in reservoir
P_{BH} Bottomhole pressure
P_{Htbg} Hydrostatic head in tubing
ΔP_{tbg} Pressure losses in tubing
ΔP_{choke} Pressure losses due to choke
$P_{Hf\ell}$ Hydrostatic head in flowlines
$\Delta P_{f\ell}$ Pressure drop in flowlines
P_{sep} Pressure at process facilities inlet

Fig. 1.5 Fluid path from the reservoir to the process facilities (*Source:* After IFP, Report No. 14048, Dec. 1966).

The bottomhole pressure from downstream to upstream is therefore the sum of the following terms:

P_{sep} pressure required at the surface treatment facility inlet
ΔP_{fl} pressure losses in the flowlines
PH_{fl} variation in hydrostatic pressure between the process facilities and the wellhead
ΔP_{choke} pressure losses at the wellhead choke (choke used to regulate the well flow rate)
ΔP_{tbg} pressure losses in the tubing between the bottom of the wellbore and the surface
P_{Htbg} variation in hydrostatic pressure between the wellhead and the bottom of the wellbore

As long as there is a choke bean at the wellhead it means that the well has a flow capacity that is greater than that which is needed or authorized at the time (because of regulatory restrictions, reservoir problems such as coning, surface processing considerations, etc.). The problem is not a lack of flow capability but one of regulations, connection between the formation and the wellbore, surface processing or dispatching capacity, etc.

Let us now consider the opposite case where the well flow rate is lower than that which is required even when there is no choke bean at the wellhead. Boosting the flow entails:

- decreasing the back pressure P_{BH}
- increasing P_R or more commonly limiting the reservoir pressure drop, a decline due to the volume that has already been produced
- increasing PI or C.

1.2.2.1 Decreasing the back pressure P_{BH}

Let us see to what extent it is possible to minimize the different terms that make up P_{BH}, excluding ΔP_{choke}.

A. *The case of an oil well*

a. P_{sep}

The separation pressure first of all affects the quality of gravity separation. The velocity of the oil droplets that are carried along by the gas depends on this pressure. As a result, depending on the amount of gas, sufficient pressure is needed so that the gas velocity is not excessively high.

The separation pressure also influences the thermodynamic efficiency of separation, i.e. the amount of liquid ultimately recovered in the storage tank after multistage separation (Fig. 1.6) for a given hydrocarbon mass coming into the separator, since the rest is "lost" in the form of gas.

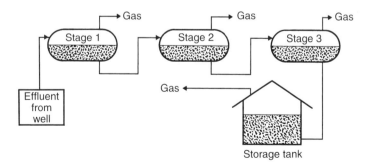

Fig. 1.6 Multistage separation: simplified diagram (*Source:* ENSPM Formation Industrie).

The pressure required in the first separation stage to achieve this optimum generally ranges between 0.3 and 2.8 MPa (50 and 400 psi), at least for effluents that do not contain too many light components.

It may be advantageous to choose a treatment pressure that is lower than this thermodynamic optimum (at the same time remaining high enough so that the gas velocity is not excessive). Thermodynamic efficiency is as a result naturally a little lower, however the increased flow from the well (particularly if the *PI* is good) usually more than offsets this drawback.

b. ΔP_{fl}

The facilities are usually designed from the outset to minimize this term. The value is commonly lower than 0.5 MPa (75 psi).

Besides the price of the flowlines, there are not normally any particular surface constraints that would prevent the use of an appropriate flowline diameter.

c. P_{Hfl}

Except in special cases, the flowline is more or less horizontal and so this term which is governed by the topology of the site is naturally very small.

d. ΔP_{tbg}

For low and average flow rates (less than a few hundred cubic meters per day or a few thousand barrels per day), conventional drilling and casing programs generally allow for enough space for the tubing so that this term is relatively small (approximately 0.5 to 1 MPa [75 to 150 psi]).

When higher flow rates are expected or for special configurations (multiple completion, etc.), a compromise must be found between the negative effect of pressure losses on the well's flow capacity and the additional drilling costs entailed by a larger wellbore diameter.

e. P_{Htbg}

This is the term that accounts for the largest proportion of bottomhole pressure. For example, it will usually range between 14 and 20 MPa for a well with 2000 m of vertical depth (between 2000 and 3000 psi for a well with 7000 ft of vertical depth). It is therefore an essential factor in determining whether an artificial lift method is to be used.

The hydrostatic pressure must be lower than the reservoir pressure for the well to be naturally flowing, if not artificial lift is necessary.

Even if the hydrostatic pressure is lower than the reservoir pressure it can still be too high for the well to flow at the required rate. This is because, on the one hand, bottomhole pressure involves other terms (in particular P_{sep}, ΔP_{fl} and ΔP_{tbg}) and on the other, the difference necessary between P_R and P_{BH} is governed by the required flow rate. If this is the case, an artificial lift process must also be implemented, since the flow rate can thereby be increased.

Accordingly, instead of making a distinction between flowing wells ($P_R > P_{Htbg}$) and non-flowing wells ($P_R < P_{Htbg}$), it is more practical to make a difference between wells that flow sufficiently (capable of flowing naturally at the required flow rate) and wells that are non-flowing or flow insufficiently (that can not flow naturally at the required rate).

The hydrostatic pressure P_{Htbg} depends of course on the depth of the well, but also on the mean density of the produced effluent. Depending on what is occurring in the reservoir (decrease of the reservoir pressure allowing the oil to lose its gas, changes in interfaces, etc.), the hydrostatic pressure changes over time with the percentage of free gas (favorable effect) or of water (unfavorable effect) associated with the oil.

The basic artificial lift methods used to activate production consist in lowering this term P_{Htbg} artificially.

Pumping, where a pump is installed in the well, causes a decrease in the height of the liquid column resting on the formation. Here, all that is really needed is for the liquid to be able to reach the pump, and if the pump is powerful enough it can provide the energy required to overcome not only what remains of the hydrostatic pressure up to the surface, but also the pressure losses downstream from the pump as well as the processing pressure.

With gas lift, gas is injected (directly into the tubing through the annulus of the producing well) in order to lessen the mean density of the produced effluent, which in turn causes a drop in hydrostatic pressure.

B. The case of a gas well

a. P_{Htbg} and P_{Hfl}

Here the mean density of the effluent is much lower and consequently the hydrostatic pressure is too. For example, a well 2000 m deep will generally have a hydrostatic pressure under 5 MPa (or 700 psi for a well 7000 ft deep), provided there is no accumulation of liquid in the well.

In addition, as the reservoir pressure becomes lower, the density of the gas also gets lower.

The problem therefore does not involve hydrostatic pressure.

b. ΔP_{tbg} and ΔP_{fl}

Once again because of low density of gas, gas wells are produced at much higher volume flow rates than oil wells. Given the drilling and casing constraints, more pressure losses must often be accepted, particularly in the tubing (2 MPa approximately [300 psi] for a tubing 2 000 m [6 600 ft] long).

c. P_{sep}

The separator pressure is of interest for a number of reasons:
- Excessively low pressure means much larger processing equipment if overly high velocity is to be prevented.
- From a thermodynamic standpoint, processing can be easier within a certain pressure range.

- And most importantly, in order for the gas to be dispatched by pipeline some pressure is needed after processing (from several tenths to over 10 MPa, or several tens to over 1500 psi, depending on the distance from the consumption node even when there are booster stations along the line).

As a result, if a gas well's reservoir pressure is inadequate to meet pressure requirements, the extra energy is not added in the well as for an oil well, but on the surface by compressors. Depending on the case, compressors are installed at the end of the process before dispatching, ahead of the processing facilities, at the wellhead or located in all these different places.

1.2.2.2 Slowing down the decline in P_R

The reservoir pressure tends to get lower as the amount of produced oil (or gas) increases and this can happen rapidly or slowly depending on the reservoir's natural drive mechanisms (whether there is a gas cap, water drive, etc.).

Water (or gas) injection wells can partly (or totally) offset the produced volume and therefore partly (or totally) maintain the reservoir pressure. In this way flow capability problems that might eventually crop up can be prevented or attenuated.

No confusion should be made between gas injection to maintain pressure (injection into the reservoir itself, far from the production well and generally into a gas cap) and gas injection to lift production from a well by gas lift (injection directly into the production well).

Maintaining pressure by injection comes under the heading of what is called secondary (or artificial) production (or recovery or drive). This is in contrast with primary (or natural) production (or recovery or drive) which covers the natural mechanisms (gas cap, aquifer, etc.). Actually, in secondary production, pressure maintenance works in conjunction with a sweeping effect whereby the oil or gas in place is flushed along by the injected fluid.

1.2.2.3 Increasing *PI* or *C*

Another way of boosting a well's delivery consists in attempting to increase the productivity index with the same pressure difference $(P_R - P_{BH})$ on the formation.

The productivity index can be insufficient either because it is naturally poor or because it was affected when the formation was drilled or when the well was completed.

The methods that can be implemented are as follows:

- Preventive care when drilling in and during completion in order to prevent or attenuate damage.
- Processes to unplug perforations or sometimes simply reperforating (when the perforated height is insufficient).
- Well stimulation methods, aimed at either unplugging the formation in the vicinity of the wellbore (when it has been damaged) or creating a drain in the reservoir (when the reservoir's permeability is low or very low) — in particular acid well treatment and fracturing, but also horizontal drilling.

- More particular processes aiming to decrease in-place fluid viscosity (injection of steam or surfactants, in situ combustion, etc.).

The last ones on the list belong to what is termed tertiary production (or recovery or drive), or improved oil recovery (IOR). Their effect on the productivity index works in conjunction with the sweeping (and pressure maintenance) effect. Note that what is known as enhanced oil recovery (EOR) covers both secondary and improved oil recovery.

1.2.3 Performance curves

For a given system (i.e. one tubing, one flowline, one separation pressure, etc.) and given reservoir conditions (i.e. one reservoir pressure, one productivity index, etc.), the maximum well flow rate can be determined by means of a graph-type representation.

a) Draw the *IPR* curve (i.e. the Inflow Performance Response curve) giving the bottomhole pressure versus the flow rate based on the reservoir pressure, i.e. for an oil (Diagram 1.2):
  (1) The part where the formula $Q = PI(P_R - P_{BH})$ applies.
  (2) The part where the formula no longer applies:
  - Because the flow in the vicinity of the wellbore is too fast (turbulence).
  - And/or because the flow is two phase in the vicinity of the wellbore (existence of free gas).

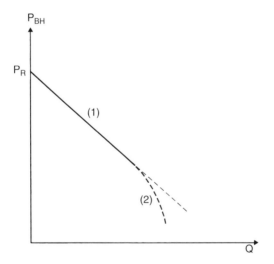

Diagram 1.2 (*Source:* ENSPM Formation Industrie).

b) On the same graph, draw the *SIP* curve (i.e. System Intake Performance curve) giving the bottomhole pressure versus the flow rate based on the downstream back pressures (Diagram 1.3):

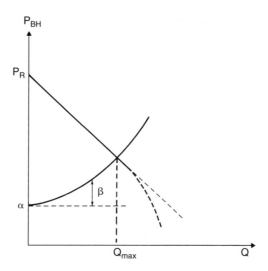

Diagram 1.3 (*Source:* ENSPM Formation Industrie).

- The ordinate at the origin (α) corresponds to the sum of hydrostatic pressures (P_{Htbg} and P_{Hfl}) and the separation pressure (P_{sep}).
- The rise of the curve (β) corresponds to the sum of the pressure losses in the tubing and in the flowline (ΔP_{tbg} and ΔP_{fl}).
- The maximum possible flow rate for the well is given by the intersection of these two curves, *IPR* and *SIP*.

Any action taken on the system (well stimulation, change in tubing diameter, change in separation pressure, etc.) means modifying one curve or the other, and therefore modifying the operating point. As a result, analyzing the curves can help find out if the modification of some parameter will have any perceived effect on the maximum well flow rate.

In addition, the pressure drop needed at the choke can be read directly on the graph for any operating flow rate lower than the maximum. It is the difference in pressure between the two curves for the flow rate under consideration.

1.2.4 Synthesis

Depending on the context, priority can be given more directly to increasing the flow rate or to the duration of the naturally flowing production phase, but these are just two aspects of the same problem.

A close interrelationship can be seen between maintaining pressure, artificial lift, well stimulation, drilling and casing program and surface processing (separation pressure, etc.). The choice of taking one or more actions can therefore only be made from the standpoint of overall field development and after assessing the costs and economic consequences.

1.3 MAJOR TYPES OF COMPLETION CONFIGURATIONS

The purpose of completion is to enable wells to be exploited as rationally and economically as possible and it can involve a large number of configurations. The technical person's job is to know how to choose the one that is best suited to the problem that needs to be solved. Good completion can not be improvised or "off the rack", here only "tailor made" can fill the bill. There is usually no ideal solution, but there are compromises and often the most economical one possible is what is chosen. Attention is called to the fact that the solution which is initially the cheapest is not necessarily the most economical in the long run, if there is a risk it will lead to costly maintenance work. However, the opposite extreme should also be avoided.

In selecting the type of completion, certain principles of relativity and anticipation must be kept in mind:

- How do completion and maintenance costs compare with expected profits? Obviously a very large field producing high quality oil with high flow rates per well warrants greater expenditure than a small one with an uncertain future that does not produce particularly commercial oil.
- How does a possible money-saving measure compare with the risks it implies? In other words, is a given risk worth taking, given the foreseeable financial consequences and the probability that something will go wrong?
- How will the production of the field and of the given well evolve in theory? The type of completion chosen must either be adapted from the outset to the way production will proceed or be capable of easy modification to meet future changes. The worst mistake, the one that must be avoided, is to wind up in a situation that has no solution.

1.3.1 Basic requirements

Depending on the case under consideration, there may be a number of requirements, and whatever the case some of them come up again and again. Special attention should be paid to them. Above all the completion configuration must be able to solve the following problems effectively:

- maintain borehole wall stability, if need be
- ensure selective production of the fluid or formation, if need be
- create a minimum amount of restrictions in the flow path
- ensure well safety
- allow the well flow rate to be adjusted
- allow operations to be performed on the well at a later date (measurements, maintenance, etc.) without having to resort to workover
- make workover easier when it does become necessary.

1.3.1.1 Borehole wall stability

This point is self-evident for wells with wall stability problems from the beginning. For some boreholes, wall stability may deteriorate with time due to various factors (depletion, water cut, etc.). Even in this case, it is important for the problem to be solved as soon as the well is brought on stream to maintain technical efficiency and avoid costly workover jobs.

1.3.1.2 Selectivity

The problem may either involve one borehole that penetrates several reservoir formations or one reservoir containing several fluids. It is necessary to understand the reservoir and its behavior over time (particularly the changes in interfaces), especially in the second case.

The contrast in mobility (ratio between permeability and viscosity for a given fluid) between the target fluid and the other fluids present is also a very important parameter. It is particularly unfavorable for oil and gas (in an oil well).

1.3.1.3 Minimizing restrictions in the flow path

In the long run all energy consumption in the form of pressure losses (whether in the reservoir itself, at the level of the pay zone-borehole connection, in the well itself, at the wellhead or in the flowlines) has a negative effect, either in terms of flow rate or natural flow capability. As a result, it is important to endeavor to minimize these restrictions.

1.3.1.4 Well safety

Here we mean both safety during completion operations as such and also safety later on during production. The main points that need to be taken into consideration are the pressure (downhole pressure control during operations to prevent blowouts, and equipment's resistance to pressure), equipment corrosion and erosion, and effluent toxicity.

1.3.1.5 Flow adjustment

During production, it must be possible to control the flow of a well. In particular, reservoir considerations or local regulations may mean that the flow rate must deliberately be limited.

1.3.1.6 Operations to be performed at a later date

A number of measurement and maintenance operations are required in order to monitor the reservoir and maintain the means of production, i.e. the well. This should be practicable without having to resort to workover. It may also be advisable to be able to carry out certain adaptations or modifications according to the changing operating conditions without having to pull out the well equipment.

However, besides the fact that changing conditions are not always easy to predict, we must also listen to reason on this point. Flexibility in completion usually goes hand in hand with a greater degree of sophistication and consequently an increased risk of malfunctions and much more difficult workover jobs (when they finally become necessary).

To the extent possible, the general design and choice of equipment must therefore be conceived of in such a way as to make workover operations easier.

1.3.2 Pay zone-borehole connection: basic configurations

There are two main types of connections between the pay zone and the borehole:
- open hole completions
- cased hole completions.

Here we will only deal with the general criteria for choosing between open hole and cased hole systems.

However, there are three essential points that should not be forgotten:
- The perforation method (and the type of perforator) used if cased hole completion is selected.
- The sand control method, should one be required.
- The well stimulation method, if the problem arises.

These methods and their respective implications, advantages and drawbacks are discussed in Chapter 2.

1.3.2.1 Open hole completions (Fig. 1.7)

The pay zone is drilled after a casing has been run in and cemented at the top of the reservoir. It is left as is and produces directly through the uncased height of the borehole. This simple solution can not solve any problems of borehole stability, or selectivity of fluid or level to be produced.

A variation on the system consists in placing a preperforated liner opposite the producing layer, thereby keeping the borehole walls from caving in (but this does not solve the sand control problem). There are some special solutions for sand control but they will not be covered here (see Chapter 2).

Open hole completions are used where there is only one zone which is either very well consolidated or provided with open-hole gravel packing for sand control. This is valid as long as there are — theoretically at least — no interface problems.

Because of this, open hole completions are seldom chosen for oil wells (a water-oil or oil-gas interface is frequently present from the beginning or later on; the oil-gas interface is even more serious due to the high mobility of gas as compared to oil). On the other hand, this type

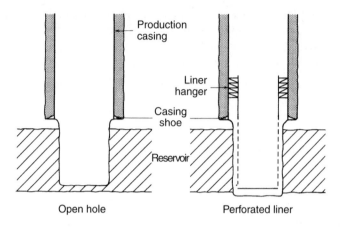

Fig. 1.7 Open hole completion (*Source:* ENSPM Formation Industrie).

of connection may be suited to a gas well. In this case, the substantial mobility contrast between the gas and the liquids is favorable and provides natural selectivity to produce mainly the gas (do not forget, however, that the accumulation of liquids in the well has a very adverse effect on the well's flow capability).

1.3.2.2 Cased hole completions (Fig. 1.8)

After the pay zone has been drilled, a casing (or a liner in some instances) is run in and cemented opposite the layer. Then it is perforated opposite the zone that is to be produced in order to restore a connection between the reservoir and the well. The perforations will have to go through the casing and the sheath of cement before they penetrate the formation. The preceding drilling phase was stopped just above the reservoir or at some distance above it and an intermediate casing was then run in and cemented.

Since perforations can be placed very accurately in relation to the different levels and interfaces between fluids, this method gives better selectivity for levels and produced fluids. The only condition, however, is a good cement bond between the formation and the casing string.

As mentioned before, the special sand control solutions will not be dealt with here. It should be pointed out, however, that a "window" may be made for sand control. This operation consists in milling the casing for a specified height, then drilling through the cement sheath and the adjacent part of the formation with a reamer.

Cased hole completions are mainly used when there are interface problems and/or when there are several levels. As a result, they are very common. It is even the one which is normally used on expansive wells in order to protect them at best from future problems.

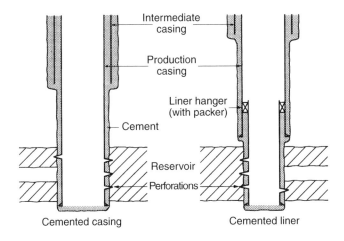

Fig. 1.8 Cased hole completion (*Source:* ENSPM Formation Industrie).

1.3.3 Main configurations of production string(s)

These configurations basically depend on the number of levels due for production and whether a production string (tubing) is used (conventional completion) or not (tubingless completion).

1.3.3.1 Conventional completions

This is a system whereby one or more production strings (tubing) are used for safety and/or other reasons. The rest of the equipment is not determined (whether there is a production packer, etc.). The fundamental characteristic of the tubing is that it is located completely inside the casing and that it is not cemented, therefore easy to replace.

A. *Single-zone conventional completions* (Fig. 1.9)

In single-zone conventional completions, the well is equipped with a single tubing to transport the production of this zone up to the surface.

There are two main types of single-zone completions, depending on whether the tubing has a production packer on its lower end. The packer provides a seal between the casing and the tubing, thereby isolating and protecting the casing. Depending on the constraints we mentioned earlier, other elements are generally included in the production string. They will not be discussed here (see Chapter 3).

1. INTRODUCTION TO COMPLETION

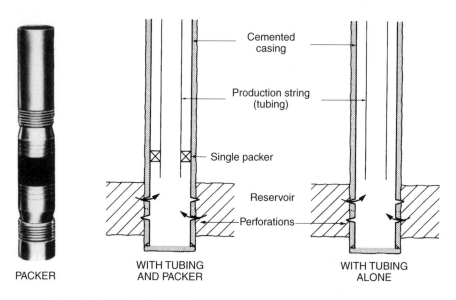

Fig. 1.9 Conventional single-zone completion (*Sources:* Baker catalog, 1984–1985, and ENSPM Formation Industrie).

Single-zone completions with just a tubing and no production packer are used when the only aim is to have the right pipe diameter with respect to the flow rate. By this we mean obtaining enough velocity to lift the heavy part of the effluent (for instance, water or condensate in a gas well) but not too much in order to limit pressure drops, thereby minimizing energy consumption: $P_{Htbg} + \Delta P_{tbg}$. They may sometimes be considered as a variation on single-zone tubingless completion, since the hanging tubing has more of a repair and maintenance function (kill string to neutralize the well for workover jobs for example). They may be suitable for wells that produce a fluid that causes no problems at a very high flow rate. The well is then produced through the tubing and the annulus.

Single-zone completions with a tubing and a production packer are the most widely used because of:

- The safety due to the packer (government regulations and company rules increasingly stipulate that a packer is to be used particularly offshore in conjunction with a subsurface safety valve on the tubing).
- Their relative simplicity in comparison with multiple or other types of completion, in terms of installation, maintenance and workover.

B. Multiple-zone conventional completions

In the past, the technique of producing several levels together through the same tubing was used. It required only a minimum amount of equipment. However, the subsequent reservoir

1. INTRODUCTION TO COMPLETION

and production problems that were experienced have caused this practice to become much less common.

a. Parallel tubing string completion and tubing-annulus completion (Fig. 1.10)

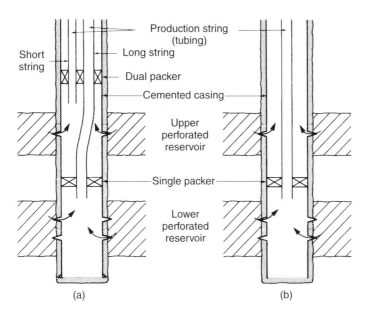

Fig. 1.10 (a) Parallel dual string completion and (b) tubing-annulus completion (*Source:* ENSPM Formation Industrie).

Here several levels are produced in the same well at the same time but separately, i.e. through different strings of pipe.

Double-zone completions are the most common, but there can be three, four and even more levels produced separately. However, this significantly complicates the equipment that needs to be run into the well and especially makes any workover operations much more complex.

There are a large number of systems, but let us simply consider:
- Parallel dual string completion with two tubings, one for each of the two levels and two packers to isolate the levels from one another and protect the annulus.
- Tubing-annulus completion with one single tubing and one packer, which is located between the two levels that are to be produced, with one level produced through the tubing and the other through the tubing-casing annulus.

Once again other elements are usually included in the production string (according to the requirements mentioned earlier), but will not be dealt with here.

Basically this type of completion allows the development of several levels with fewer wells, and is therefore faster. In contrast, maintenance and workover costs are higher. As such it is particularly advantageous offshore (where drilling itself and the space required for a well site are very costly). By taking advantage of a principal level that is being produced, it is also used to develop one or more marginal level(s) that would otherwise not warrant drilling a well.

However, it should be borne in mind that the ideal completion is the simplest. It will entail the simplest operations in terms of installation, maintenance and workover.

Tubing-annulus completions are very few and far between. Though they have good flow capability (large cross-sections are available for fluid flow), this system does not protect the casing, among other drawbacks.

Parallel dual tubing string completions are therefore the typical textbook example of multiple-zone completion. More sophisticated completions require careful study in order to avoid:
- problems in operation and production due to frequent wireline jobs
- problems of safety and operation during workover.

b. Alternate selective completions (Fig. 1.11)

Here the idea is to produce several levels in the same well separately but one after the other through the same tubing without having to resort to workover. Production alternates in fact and wireline techniques are used to change levels.

This type of completion is especially suited to a situation where one of the two levels is a secondary objective (very rapid depletion, simple observation from time to time, etc.) which would not warrant drilling a well.

Besides packers, this technology also requires extra downhole equipment such as:
- a circulating device consisting of a sliding sleeve to open or obstruct communication ports between the inside of the tubing and the annulus
- a landing nipple allowing a plug to be set in the well.

Parallel tubing string and alternate selective completion systems can be combined. For example two parallel tubings, each equipped for two levels in an alternate selective manner, can produce four levels separately, provided that only two are produced at the same time.

1.3.3.2 Tubingless completions

A tubingless completion uses no tubing, but production flows through a cemented pipe instead. This rather unusual type of completion, which will be covered only briefly here, is mainly used in certain regions and only under specific conditions.

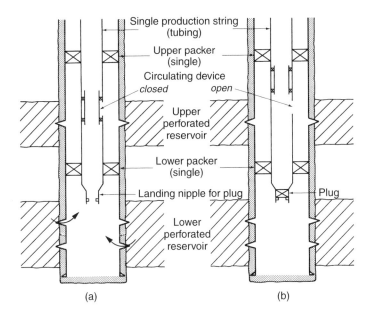

Fig. 1.11 Alternate selective completion (*Source:* ENSPM Formation Industrie).

A. *Single-zone tubingless completions* (Fig. 1.12a)

Production flows directly through a casing, usually of large diameter. Wells that are big producers of trouble-free fluids can be exploited in this way with minimum pressure losses and the lowest possible initial investment. This system is found particularly in the Middle East.

B. *Multiple-zone tubingless completions* (Fig. 1.12b)

Production flows directly through several casings whose diameters may be very different from one another depending on the production expected from each level.

Several levels with mediocre production can be produced in this way with a minimum number of wells and downhole equipment, i.e. with a minimum initial investment. This is true provided there are no safety or production problems (artificial lift, workover, etc.). This type of completion is mainly encountered in the United States.

1.3.3.3 Miniaturized completions

This system basically involves multiple-zone tubingless completions equipped with little macaroni tubings so that each cemented casing has a conventional single, or more often tubing-annulus completion. It is of course highly specific and is found in the United States as is the previous type.

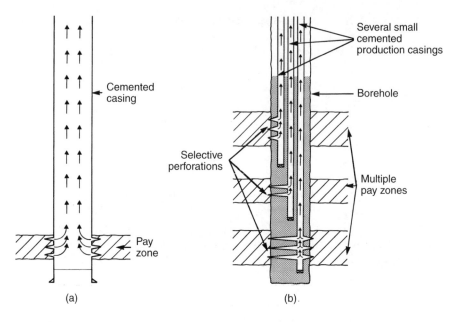

Fig. 1.12 Tubingless completion (*Source:* ENSPM Formation Industrie).
a. Single zone. b. Multiple zone.

Another case can also be mentioned: that of slim-hole drilled wells that are equipped with a tubing (such as those drilled in 1982 on a field in the Paris basin, even though they are produced with a tubingless completion system through a production casing: some 50 mm (2") in diameter at nearly 1500 m (4400 ft) depth with an original hole diameter of about 153 mm (6")).

1.4 MAIN PHASES IN COMPLETION

Depending on the context, well completion may involve different phases and the order of the phases can vary. The procedure below is only one among many possible solutions since certain phases can be carried out at other times or may prove unnecessary.

Bearing in mind that the drilling in conditions are crucial for completion, we consider here that the pay zone has already been drilled, that the open hole logging has been done and that the casing has also already been run in and cemented if the completion is cased hole.

1.4.1 Checking and conditioning the borehole

This operation consists in checking the conditions of the borehole. In an open hole this usually means simply running the drill bit back in hole to the bottom and circulating in order to get uniform mud. Sometimes the mud that was used to drill the layer can be replaced by a completion fluid that is more appropriate for further operations.

For a cased hole the operations are more numerous:
- A drill string equipped with a bit and a scraper are run in. In order to do so, the drill collars and part or all of the drill pipe used to drill the well may have to be laid down. Then another string with smaller diameter components is run in.
- The top of the cement in the casing is checked with the bit (and any excess cement is redrilled).
- The area where the packer is to be set is scraped, mud is circulated at the same time to get rid of the cuttings (particles of cement and other debris).
- Completion fluid is pumped into the well at the end of circulation and the drill string is pulled out.
- Logs are run to check the cement bond quality and for all other reservoir engineering needs.
- Depth checking logs are run. These are usually a gamma ray along with a CCL (Casing Collar Locator). The gamma ray (γR) gives a correlation with the logs that were run in the open hole (which generally included a gamma ray) and the CCL records the casing couplings. Afterwards the tools used during completion operations, for example perforating guns, can be depth matched with respect to reservoir levels or interfaces simply on the basis of casing collars, or directly on the basis of γR.

1.4.2 Remedial cementing

If the cement bond quality is inadequate to handle reservoir problems (isolation of levels, interfaces, etc.) and operations on the pay zone (well testing, treatment, etc.), remedial cementing becomes necessary. Here the casing is usually perforated and cement is injected under pressure opposite the area(s) that were improperly cemented.

1.4.3 Re-establishing pay zone-borehole communication

In a cased hole, the barrier consisting of the casing and the sheath of cement between the reservoir and the wellbore must be bypassed. Except for special cases this is done before or after equipment installation by perforating with shaped charges. This operation entails extremely stringent safety practices related to the use of explosives. In addition, when the well is perforated before running in the production string, care must be taken to prevent the well from flowing from this point on, until the permanent equipment has been installed.

1.4.4 Well testing

All wells must be tested, sometimes for a short time, in order to ascertain the productivity (or injectivity) index and any possible damage.

For a development well in an area where many other wells have already been completed, a simple assessment at the end of completion may sometimes be enough.

In contrast, for the first development wells — and even more so for appraisal and exploration wells — more thorough testing is needed. Based on these tests and further rock and fluid studies in the laboratory the need for treatment and the suitability of one treatment over others can be determined.

Well testing is therefore quite often performed with temporary assemblies.

1.4.5 Treating the pay zone

Well treatments mainly encompass sand control operations and well stimulation (acidizing, fracturing). Well testing may be necessary prior to well stimulation to establish whether the operation is worthwhile. Depending on the case, treatments are carried out before or after equipment installation and may require temporary equipment.

1.4.6 Equipment installation

Here the permanent equipment is installed in the well and is then tested. As we have already mentioned, the equipment can be installed before or after the well is perforated.

The basic conventional equipment (packer, miscellaneous downhole equipment, tubing, wellhead) may also be supplemented by specific safety or artificial lift equipment.

1.4.7 Putting the well on stream and assessing performance

For a production well this phase requires the following: the original fluid whose density was high enough to counter the reservoir pressure prevailing at the outset in the well must be replaced by a lighter one. Depending on when the perforations are made (before or after equipment installation) and according to the equipment that is installed, this operation can take place at different times (after equipment installation or before perforation).

Putting the well on stream also includes a clearing, or unloading phase. Treatments carried out on the layer may entail constraints during this phase (start clearing as soon as possible after acidizing, gradual clearing in the event of sand control, etc.). Assisted start up may prove necessary (swabbing, using a coiled tubing, etc.), not to speak of permanent artificial lift processes.

Initial assessment of well performance is essential. It can serve to decide whether a treatment is suitable and is used as a reference when the well's behavior is monitored later on. Lastly it may provide a wealth of information of use for future wells.

1.4.8 Moving the rig

When the drilling or completion rig is moved, the well must always be placed under safety conditions. The rig can be moved once all of the operations have been concluded or as soon as the permanent equipment has been installed.

1.4.9 Operations to be performed at a later date: measurements, maintenance, workover and abandonment

The well's behavior is monitored as time goes by and action is taken accordingly. The decisions that were made can be checked to see if they were justified and how well they were implemented on the site. The reservoir's behavior can also be better understood and therefore production problems can be anticipated.

Additionally, the well requires maintenance to keep it producing properly. It may also undergo workover for repairs or to modify production conditions.

In order to do this the following information is essential:

- all the operations that were carried in the well and the conditions in which they were performed
- all of the equipment installed and its characteristics.

Consequently, a detailed report should be drafted at each step and all of the information can be entered into a computerized data bank.

Chapter 2

CONNECTING THE PAY ZONE AND THE BOREHOLE

The main factors that influence the way the pay zone-borehole connection is designed and the principal configurations of the connection were discussed in the preceding chapter.

The present chapter covers:

- the different phases required to make the connection
- the techniques that can be used and their implementation.

2.1 DRILLING AND CASING THE PAY ZONE

Without going into considerations related purely to drilling, we will discuss only a few particularly important points for completion.

2.1.1 Well safety

The pay zone that is going to be produced is by nature porous, permeable and contains fluids under pressure. This means that the conditions for a blowout are all present.

As such, it is critical to be sure that the drilling fluid in the well has sufficient density to counter the reservoir pressure before starting this phase. It is also essential to check that safety equipment, especially the BOPs (blowout preventers), are in good condition and operate properly, since the whole structure was designed (drilling and casing program) so that the well would withstand the reservoir pressure.

Likewise, during all operations on the well, great care must be taken to keep kicks from occurring. Swabbing needs particular attention when the string is being pulled out. Another point to remember is that measures implemented to control a kick might damage the pay zone and affect its productivity characteristics.

2.1.2 Fluids used to drill in the pay zone

2.1.2.1 Constraints

A. Safety constraints

As mentioned earlier, the fluid in the well must exert higher hydrostatic pressure than the reservoir pressure. In practice, at least for development wells, the excess pressure is generally set at around 0.5 to 15 MPa. Either a fluid loaded with solids or brine is used to get enough specific gravity.

When the required specific gravity is being determined, be careful to:
- use a system of consistent units (in particular do not directly compare a reservoir pressure in bar with a hydrostatic pressure expressed in kgf/cm^2)
- pay due attention to the effect of temperature on the mean fluid specific gravity (temperature can cause the mean specific gravity in the well to decrease by several percentage points).

B. Drilling constraints

Since the fluid is used for drilling, it must have the properties that are normally required for a drilling fluid. In particular, it must have sufficient viscosity to lift the cuttings up out of the borehole.

C. Formation damage constraints

Because of the excess hydrostatic pressure, the fluid in the well tends to flow into the reservoir. Although the liquid part of the fluid (the filtrate) can in fact invade the formation easily, the solids contained in the fluid tend to remain on the borehole wall where they form a deposit called the filter cake.

In practice the most common types of damage are:
- Formation plugging due to the solid particles in the drilling fluid that penetrate the formation. Because of the pore diameter, plugging is usually fairly shallow and we can hope to solve the problem if the particles manage to come back out again when the well is put on stream or if they are soluble in acid (but this is in no way always true).
- Plugging due to the action of the filtrate on the matrix itself and in particular on shale formations (damage by swelling and dispersion of shales), or the effect on formation wettability with the consequences on relative permeability to oil and to water.
- Plugging due to the action of the filtrate on the fluids in place in the reservoir, in particular problems of precipitates, emulsion or sludge (a flocculated mass of heavy oil).

It should be emphasized that these cases of damage are usually irreversible.

The filter cake that is formed on the borehole wall is not particularly troublesome, at least not in production wells. Here the flow from the formation will tend to slough the cake off the wall. In an injection well however, it will lower injectivity considerably if it has not been removed. On the contrary, the cake limits filtration, thereby reducing damage to a producing layer.

The effect of plugging in the vicinity of the wellbore on the productivity index is illustrated on the graph in Figure 2.1. Even very shallow damage can be seen to cause a substantial reduction in productivity.

2. CONNECTING THE PAY ZONE AND THE BOREHOLE

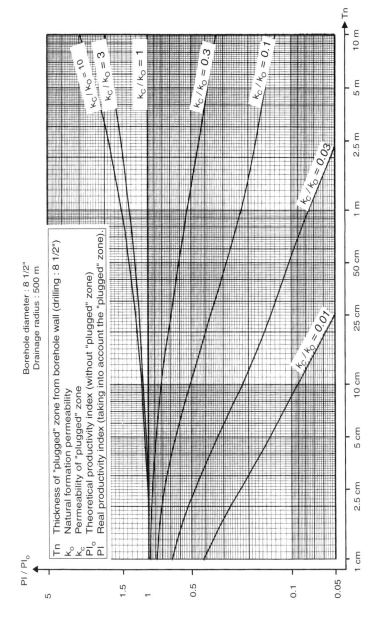

Fig. 2.1 Influence on the productivity index of variations in permeability in the vicinity of the wellbore (in radial flow) (*Source:* ENSPM Formation Industrie).

For formations that react well to acid (usually true for carbonates) and if the thickness of the plugged zone is small, this is not too serious, since plugging can be bypassed afterwards by attacking the formation itself with acid. However, for those that do not react well (usually true for sandstones), since neither formation nor plugging agents react the damage may well be permanent. In addition, side reactions may make plugging worse.

Consequently in these cases prevention is essential and it is crucial to use a suitable drilling fluid as soon as the pay zone is entered. It may also be advantageous to have a high drilling speed and a low overpressure on the pay zone to limit the amount of filtrate that invades the formation.

For damage, the ideal situation would be to have in the well a fluid with no solids, whose filtrate would be compatible with the formation and the fluids in place in the reservoir. However, it is generally necessary to add what is called temporary plugging agents to the drilling fluid (they are either soluble in acid or unstable with temperature but they are never 100% soluble) to reduce filtration. Moreover, when the fluid is used to drill in the pay zone, viscosity agents are needed to give it sufficient lifting capacity to transport the cuttings.

When a fluid containing solids is used, preference should go to solids that are risk-free (not barite, for example) and can be "readily" broken down (acidized).

Lastly while the pay zone is being drilled in, the drilling fluid must be treated in order to remove fine-grained solids. These are produced when the solids in the fluid (originally in the fluid or formation cuttings) are ground down.

2.1.2.2 Completion fluids

This is the term for the specific fluid that is used opposite the pay zone, it is designed to cause the least damage possible to the reservoir. As discussed earlier, it must be pumped into the well before the formation is penetrated, whatever the configuration chosen for the connection between the borehole and the pay zone. This is particularly true and important for sandstone type formations which do not react well to acid. In contrast, a little more leeway is possible with other formations, especially carbonates.

It is often difficult to formulate a fluid which:
- does not damage the reservoir
- provides good characteristics with respect to drilling.

The completion fluid is therefore used mainly:
- if possible or necessary as soon as the pay zone is drilled in
- during initial completion
- to control the well (kill the well) in order to proceed to some particular treatment
- during workover, after the well has been produced, to repair or modify the well.

A. *Characteristics required for completion fluids*

Based on the main requirements (making sure the well is safe, preventing damage to the well, cleaning the well) the following characteristics are the most important.

a. Specific gravity

Specific gravity is designed to keep the well stable by exerting sufficient back pressure on the reservoir. A differential pressure of approximately 1 MPa (150 psi) between the hydrostatic and the formation pressures is often adopted to maintain a certain degree of safety while minimizing invasion of the reservoir. However, in some special cases the differential may be much lower, even close to zero. The specific gravity must be easy to adjust.

b. Viscosity

Viscosity must be sufficient to scavenge out the borehole properly, keep solids in suspension (cuttings and weighting materials) and keep gas kicks at bay.

c. Filtration rate

Solid particles must be kept from migrating into the formation pore system. To achieve this, temporary fluid-loss additives can be used whose grain size is adapted to the pore diameter. In some cases (sensitive formations in particular), it is important to keep the volume of filtrate invading the formation to a minimum.

d. Compatibility

All physicochemical reactions between the filtrate and the formation must be prevented by adapting its composition to the reservoir water (and even to the injection water in an instance of pressure maintenance) and to the sensitive components in the formation (especially shales).

e. Stability

The fluid must exhibit good stability with time and mainly be able to withstand the reservoir temperature.

f. Preparation and handling

The fluid must be fairly easy to prepare since rigs, particularly servicing rigs, are not always properly equipped. It must be neither toxic, pollutant, nor corrosive.

g. Price

Last but not least, the fluid should have the lowest possible cost price.

B. The main fluids (Table 2.1)

a. Low specific gravity

These are especially foams and muds containing oil such as:
- oil-base mud
- inverted emulsion mud (10 to 50% dispersed water)
- direct emulsion mud (20 to 45% dispersed oil).

b. Specific gravity greater than 1

Without solids

These consist of brines (water base plus salt). It should be noted that the crystallization point of brines depends on the type of salt(s) and the concentration (it can be greater than 0°C, or 32°F). The price increases exponentially with specific gravity.

Low solids content

These are also brines with fluid-loss additives and viscosifiers (of the type that can be acidized).

High solids content

These are water and viscosifier-base fluids with added weighting materials such as calcium carbonate or iron carbonate to achieve high specific gravity. Their drawbacks are the sedimentation risk and their high viscosity, but they are less expensive than high gravity brines.

Table 2.1 Major completion fluids on the market.

Foam	
0.20 to 0.30	dense foam
Oil base	
0.80 to 0.90	diesel or crude
0.85 to 0.95	oil-base or inverted emulsion mud
0.85 to 1	direct emulsion mud
Water base without solids (1)	
1 to 1.03	water - seawater - brackish water
1 to 1.16	fresh water + KCl
1 to 1.20	fresh water + NaCl
1 to 1.30	fresh water + $MgCl_2$
1 to 1.40	fresh water + $CaCl_2$
1.16 to 1.20	fresh water + KCl + NaCl
1.20 to 1.27	fresh water + NaCl + Na_2CO_3
1.20 to 1.40	fresh water + NaCl + $CaCl_2$
1.20 to 1.51	fresh water + NaCl + NaBr
1.40 to 1.70	fresh water + $CaCl_2$ + $CaBr_2$
1.70 to 1.80	fresh water + $CaBr_2$
1.80 to 2.30	fresh water + $CaBr_2$ + $ZnBr_2$
Water base plus solids	
1 to 1.70	fresh water + $CaCO_3$
1 to 1.80	fresh water + $FeCO_3$ (siderite)
1 to 1.80	drilling mud + $CaCO_3$ or $FeCO_3$
1 to 2.30	drilling mud + barite
1 to 2.30	fresh water + resins
1 to 2.30	oil-base or inverted emulsion or direct emulsion mud

(1) Pay attention to the crystallization point, especially with mixtures.

C. Additives

It is usually necessary to add a number of products such as:
- viscosifiers
- defoamers
- fluid-loss control agent
- emulsifiers (mud containing oil)
- weighting materials
- anticorrosion additives (bactericides, antioxidants, etc.).

2.1.3 Drilling and casing diameters

The graph in Fig. 2.1 (see above) shows that the productivity index increases little when the drilling diameter increases (consider infinite permeability in the vicinity of the wellbore). The diameter would have to be increased considerably to make any difference and this is not technically or economically possible in drilling.

The drilling diameter is in fact important with respect to the equipment that will be installed in the well, e.g.:
- a suitable sand control system when needed
- a tubing with sufficient diameter for the pressure losses to be acceptable
- several tubings if there are several layers to be produced separately
- specific equipment such as a packer (downhole seal between the casing and the tubing), subsurface safety valve, side pocket for a gas-lift valve (process that lightens the oil column in the tubing by injecting gas into it), centrifugal pump, etc.

2.1.4 Casing and cementing

Let us point out that the most common configuration for the pay zone-borehole connection is the cased hole. Since there is a cemented casing in this solution, selectivity of a level and/or a produced fluid is more efficient. However, this requires the cement bond to provide a good seal.

In addition, to save room for the downhole equipment it may be necessary to use a liner. This involves problems related to liner installation and proper cementing.

As far as primary cementing of a production casing is concerned, we will restrict our discussion to a reminder of the main objectives, since the methods of monitoring and restoring the cement job when needed will be dealt with in the following section.

The main objectives of primary cementing for a production casing are as follows:
- make it possible to select the fluid or the level to be produced, prevent fluids from migrating between the different corresponding zones or make treatment easier

- keep borehole walls from caving in (along with the casing)
- protect the casing, particularly from any corrosive fluids in the formations that have been drilled through.

For a production casing, the cement sheath must therefore provide a good sealing.

2.2 EVALUATING AND RESTORING THE CEMENT JOB

2.2.1 Evaluating the cement job

2.2.1.1 Major flaws encountered after primary cementing

Because of the difficulty in cementing a casing and even more so, a liner, a large number of flaws can often be found. Here we will mention only the major ones and the most frequent causes.

A. Inadequate filling

This can be due to:

- incorrect estimate of volume (caved hole, etc.)
- losses when the slurry is displaced
- unexpected setting that interrupts the operation (bad products, mixing in the wrong proportions, overly rapid dehydration, etc.).

B. Inadequate seal and/or strength

This can be due to:

- insufficient distance between the float collar and the shoe
- excessive displacement
- incomplete displacement of the mud by the slurry because the casing was not properly centered, the pumping rate was inappropriate, the spacer was not suited to the job, etc. (transition zone, fingering, gelled mud, mud cake on the formation, microannulus, etc.)
- a gas kick (loss of hydrostatic pressure when the slurry was setting, etc.)
- the slurry not setting or only partially setting (insufficient temperature in the upper part)
- poor quality slurry (product quality or quantity, mixing water, slurry mixing, etc.)
- deterioration with time (temperature, type of fluids in place, impacts, pressure variations, etc.).

2.2.1.2 Principal methods of evaluating a cement job

The quality of the production casing cement job is crucial in producing the well, so it is essential to be sure that the cement job meets the required standards. However, since the quality of the job varies as time goes by, the results of an evaluation operation are valid only for the time when the job was evaluated.

The conditions that prevailed when the cement job was done are already a good indication. In particular any irregularity should raise doubts about the quality of the job. However, an absence of irregularities in the way the job was done is not enough to qualify it as satisfactory, unless the reservoir has been studied to a certain extent and there is a sufficient statistical basis.

A. Direct evaluation

Pressure tests and especially negative pressure tests can be taken as indications of a good seal for a liner head or for perforations that were used for remedial cementing. It should be emphasized that pressure tests are not particularly significant for situations at a later date when negative pressure will prevail in the well (during production), since the filter cake acts as a one-way valve.

In addition, even negative pressure tests are highly debatable as an indication of the seal behind the casing itself. They only show whether the liner head or the remedial cementing perforations have been properly cemented. The tests do not guarantee that there will be no inflow of unwanted fluids or that there will be no interference between levels during well testing or normal production.

B. Indirect evaluation

Here we will deal with evaluation through logging.

Temperature logs simply give an indication of how far the cement has backed up in the annulus and therefore of the fact that the different zones or levels have actually been covered up. Cement hardening is an exothermic reaction. As a result, there is a higher temperature opposite cemented zones when the cement is setting than what would be expected from the normal geothermal gradient. However, this does not tell much about how completely the annulus has been filled up and how good the cement bond is, i.e. about the quality of the cement job.

Acoustic logs (and mainly logs with signals that are emitted and received) can be used to get a better picture. The time, frequency and amplitude of the sound waves are measured when they come back to the tool after a signal has been emitted. These measurements are interpreted according to a large number of parameters (shape of the borehole, position of the casing, type of formation, cake, mud and cement, etc.).

Therefore, interpretation is not an easy job. Depending on the type of tool it can give:
- an overall idea of the cement-bond quality
- the compressive strength and an image of the cement sheath.

The most common types of acoustic logs are:
- Schlumberger's CBL-VDL (Cement Bond Log-Variable Density Log)
- Schlumberger's CET (Cement Evaluation Tool).

a. The CBL-VDL (Cement Bond Log-Variable Density Log)

In this tool (Fig. 2.2a), the transmitter generates a low-frequency (20 kHz) acoustic wave train over a short period of time. The signal is propagated through the casing, the cement and the formation before it gets back to the receivers, one of which is located three feet (approximately 0.9 m) away and the other five feet (approximately 1.5 m) away from the transmitter. What is measured is therefore the longitudinal (vertical) path of the sound wave.

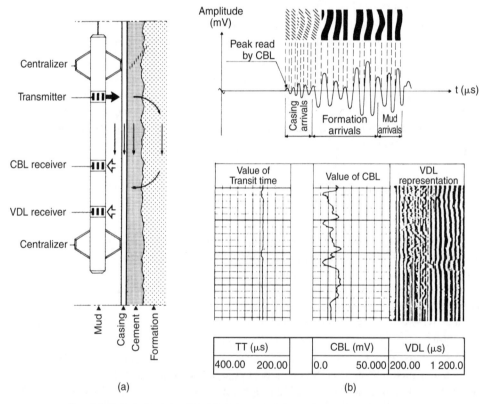

Fig. 2.2 Principle of the CBL-VDL and a standard presentation of a recording (*Source:* After *Revue de l'Institut Français du Pétrole*, 41, No. 3, 1986).

The CBL is a recording of the amplitude and the transit time of the first sound wave that reaches the receiver that is three feet away. The VDL is optional, it gives further information to supplement the CBL. The complete wave train that reaches the receiver is represented on a film by light and dark stripes, with the contrast depending on the amplitude of the positive peaks (Fig. 2.2b).

If the cement job is good, the amplitude measured by the CBL is low. The VDL shows an attenuated signal that has travelled through the casing and a strong signal that has travelled through the formation. In contrast, when the cement job is poor, the amplitude measured by the CBL is strong and the VDL shows a signal that has travelled through the casing in the form of regular, highly contrasted stripes.

Care is required since a large number of parameters (such as a coupling, a logging tool that is improperly centered in the casing, fingering or a microannulus, etc.) can affect the amplitude measurement. An attentive study of the transit time and the overall VDL picture can help take these parameters into account for interpretation. Additionally the tool provides only an idea of the average bond quality. This depends on the percentage of the circumference where the cement has bonded and the compressive strength of the cement (so the log should not be run until the cement has enough time to set).

b. The CET (Cement Evaluation Tool)

This tool has got eight high-frequency transmitter-receivers (500 kHz) located on a horizontal plane. This time the attenuation of the casing's radial (horizontal) resonance is analyzed according to the eight sectors. The attenuation depends on whether cement is present in the sector and on its compressive strength.

The tool provides three main recordings (Fig. 2.3):

- the average casing diameter and whether it is out of round
- the maximum and minimum compressive strength (averaged out over a height of four feet and a horizontal sector of 45°)
- an "image" of the cement sheath.

The image consists of eight traces (one for each sensor) varying between lighter and darker depending on the compressive strength of the cement in each sector. Shades range from white (no cement) to black (good cement). In order to improve the presentation, shading is gradually interpolated between the tracks corresponding to the different sensors (see Fig. 2.3).

The tool also has the asset of being insensitive to any microannulus less than 0.1 mm thick. On the other hand it does not characterize the cement-formation bond quality as well.

2.2.2 Remedial cementing

2.2.2.1 Introduction

The basic method used to restore an inadequate primary cement job consists in positioning cement slurry in/on the defective zone:

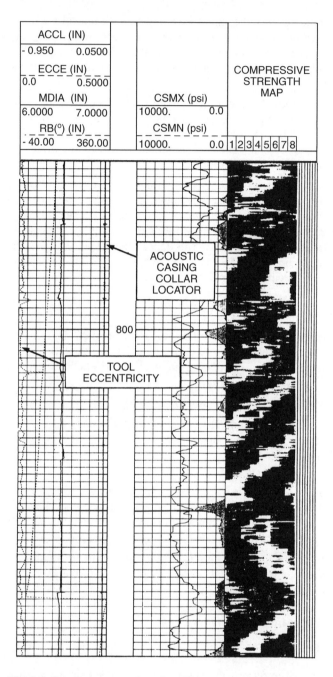

Fig. 2.3 Standard presentation of a CET recording (*Source:* Schlumberger document).

- possibly by circulating (when the problem is incomplete filling for example)
- but more commonly by pumping under pressure (squeeze), with the slurry subjected to differential pressure.

It is the differential pressure that allows the slurry to cover the defective zone by the process of the liquid base filtering into the formation. A cement cake is thereby formed on the wall. As such it is necessary to:
- monitor the growth of the cake
- allow it to form evenly on all of the permeable areas.

In addition to restoring a primary cement job, the method can also be used in workover in order to:
- restore isolation between zones
- reduce the proportion of water or gas produced due to coning
- isolate a water or gas zone
- abandon (temporarily or permanently) a pay zone
- repair a leaky casing.

2.2.2.2 Squeeze techniques

In practice, due consideration must be given to the limitations inherent to the combination between:
- the injection technique (squeeze) chosen
- the string used and the procedure involved in positioning the slurry on the zone to be treated (see Section 2.2.2.3).

A. High-pressure squeeze

The principle of this technique is to deliberately fracture the zone to be treated behind the casing so that the slurry can pass through the perforations and fill up the existing annular voids.

However, a large number of drawbacks are involved:
- The position and direction of the fracture are not controlled.
- The cement entering the fracture is "lost" since it does not contribute to the hydraulic seal between the casing and the formation.
- It is difficult to seal the fracture all throughout its extension: the displaced mud is present in front of the cement.

As a result, low-pressure squeeze is usually preferred (see following section), though high-pressure squeeze still has a number of applications:
- channels behind the casing that do not communicate directly with the perforations
- a microannulus that is impervious to slurry but not to gas
- slurry hydrostatic pressure causing fracturing
- fluids loaded with plugging solids present in the well (that have not been eliminated).

B. Low-pressure squeeze

The technique consists in forcing the slurry into spaces that communicate with a permeable formation (voids with no exit can not be filled with cement), but with a pressure lower than the fracture pressure, in other words usually at low flow rate. The water alone is forced out of the slurry, with the cement particles forming a cake.

This is possible only if the perforations and channels are free of plugging fluids (which require prior removal) and if the formation is permeable enough. Consequently an injectivity test will have to be run beforehand to be sure that the operation can be performed with a downhole injection pressure lower than the fracture pressure.

It should be noted that the downhole injection pressure (pressure applied to the slurry at the bottom of the well in order to force it through the perforations) corresponds to the surface pressure plus the hydrostatic pressure of the fluids present in the well minus the pressure losses in the well (a negligible term when the injection flow rate is low). The variation in pressure on the surface during injection is therefore not representative of the variation downhole (in particular due to the decreasing height of slurry in the well as it is pumped out of the string).

Sometimes pumping is continuous but more commonly it is on a stop and start basis (hesitation squeeze Fig. 2.4). In this connection:

- A small volume (50 to 300 liters, or 10 to 100 gallons) of slurry is pumped in and the corresponding pressure increase is noted.
- Pumping is halted for a specified time (10 to 15 minutes) and the pressure drop caused by the slurry entering the formation is noted (during this time the slurry is losing water).
- The cycle is repeated x times.

In this way pumping is easier (it is difficult to pump continuously at a sufficiently low flow rate so that the fracture pressure is not exceeded) and there is more time for the cement cake to develop.

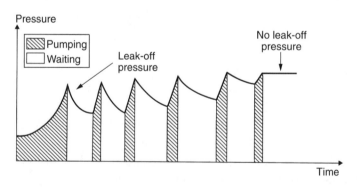

Fig. 2.4 Pressure vs time diagram (during hesitation squeeze) (*Source:* D. Veillon, *Bases techniques de la complétion,* ENSPM course material).

Hesitation squeeze is however dangerous if there is still any slurry in the string. If the slurry thickening time is not long enough, the cement may set inside the pipe.

2.2.2.3 Squeeze procedures and corresponding tool strings

In order to carry out a squeeze operation:
- the slurry must be placed at a specified point in the well
- this zone must be isolated mechanically
- so that the slurry can be injected under pressure.

Drillpipes are used to do this:
- alone, or
- with a squeeze packer: a retrievable type packer run in directly on the drill string featuring a bypass and a temporary setting mechanism (Fig. 2.5), or
- with a cement retainer: a permanent drillable packer that is not incorporated in the drill string and is equipped with a check valve to prevent backflow at the end of the operation (Fig. 2.6), and
- possibly a retrievable or permanent bridge plug, a "blind" cement retainer (Fig. 2.7).

There are three main types of methods in remedial cementing:
- slurry squeeze with the slurry displaced to perforation depth by circulating
- slurry squeeze with the slurry displaced to perforation depth by circulating then squeezing
- slurry injection while circulating.

A. *Slurry squeeze with slurry displaced to perforation depth by circulating*

The procedure is basically as follows (Fig. 2.8):
- Run in the string until the tail pipe (extension under the packer) or the drillpipes alone are below the perforations.
- Circulate the slurry until it is opposite the perforations (preferably with the BOPs closed and choking the returns to prevent the slurry from falling and avoid air pockets) (Fig. 2.8a).
- Pull up the tail pipe or the drillpipes alone above the slug of cement.
- Close the BOPs (if this has not already been done).
- Reverse circulate to clean out the pipes (preferably before squeezing) (Fig. 2.8b).
- According to the type of string, set the packer.
- Squeeze according to the chosen injection technique (Fig. 2.8c).
- Open the BOPs again and, according to the type of string, unseat the packer.
- Pull the string out.
- Run the string back in with a drilling bit and a scraper.
- If necessary, wait for the cement to set.
- Drill out the excess cement and clean the casing.

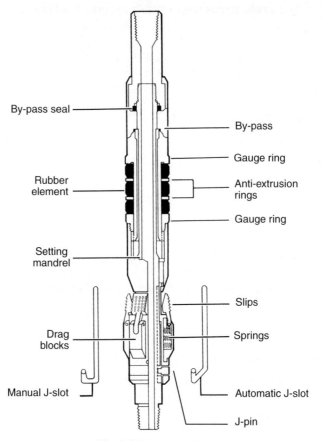

Fig. 2.5 Squeeze packer
(*Source:* Schlumberger document).

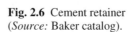

Fig. 2.6 Cement retainer
(*Source:* Baker catalog).

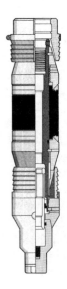

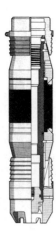

Fig. 2.7 Bridge plug
(*Source:* Baker catalog).

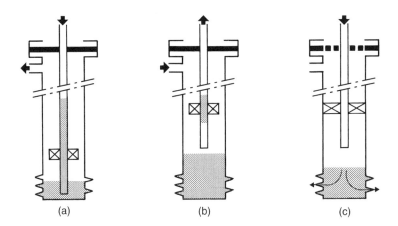

Fig. 2.8 Squeeze with packer and long tail pipe (or drillpipes alone) (*Source:* ENSPM Formation Industrie).

The advantage of these methods is that they allow the slurry to be pumped down to perforation depth before squeezing. Therefore no well fluid has to be injected into the formation to perform the squeeze.

When a packer is used:

- Use an extension under the packer (made of tubing) that is long enough to place the slurry at the bottom of the perforations by circulation without letting the slurry back up to packer depth.
- It is possible to pressurize the annulus a little depending on the planned squeeze pressure.

Using a packer means that:

- Pressure is not restricted by the wellhead or the casing above the packer.
- The zone to be treated can be isolated from perforations located at a sufficient distance above it (but the packer is set relatively far above the perforations that will be treated during the squeeze operation).

The use of drillpipes alone, a less costly technique in terms of equipment, is restricted to low pressure applications when there is no uncertainty as to the state of the casing or the injection pressure that will be required. It is most commonly used when all that is involved is placing a cement slug in the well.

B. Slurry squeeze with slurry displaced to perforation depth by circulating then squeezing

Here drillpipes are used in conjunction with a squeeze packer and a short tail pipe or a cement retainer. The procedure in this case changes significantly (Fig. 2.9):

- Run in the string.
- Set the packer as close as possible to the perforations (based on CBL results, etc.) with the tail pipe extending the string to the top of the perforations.
- Circulate the slurry as deep as possible in the string through the packer bypass (keep some safety margin), choking (mandatory) returns (Fig. 2.9a).
- Close the bypass.
- Pressurize the annulus moderately, if necessary.
- Perform the squeeze (Fig. 2.9b).
- Unseat the packer and pick it up a few meters.
- Reverse circulate (Fig. 2.9c).
- Reset the packer one stand higher if necessary and keep the well under pressure.
- Pull the string out and continue as in the first method.

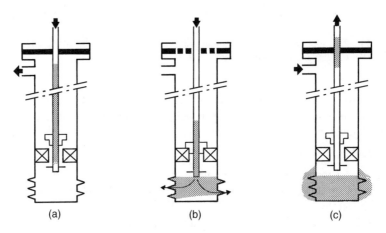

Fig. 2.9 Squeeze with packer and short tail pipe (or cement retainer) (*Source:* ENSPM Formation Industrie).

When a cement retainer is used:

- Generally set beforehand, it is not incorporated in the string and has to be drilled out to be removed.
- In order to circulate or squeeze, a stinger (a small-diameter tube that fits into a receptacle and will not form or will form a seal depending on its position) must be disengaged from or engaged in the cement retainer.

These methods are used to isolate a maximum height of casing from the squeeze pressure. In contrast, the slurry is not yet in front of the perforations at the beginning of the squeeze and so:

- Some well fluid is injected into the formation.
- The lower perforations may not be treated very well.
- There is a danger of bridging opposite the upper perforations if massive water loss occurs (consequently, do not start squeezing with a hesitation squeeze).

When a packer is used with a short tail pipe, there is a risk of stuck string if the slurry backs up to the tail pipe or packer depth (during circulation or due to a leaky packer, etc.).

There is less risk of getting stuck with a cement retainer. In addition, if it is wireline set, it can be positioned very accurately in relation to the formation by using the same technique as for perforation depth adjustment (see Section 2.3.4.2) and so it is possible to isolate two neighboring perforated zones. The cement retainer must, however, be drilled out if access to the treated zone becomes necessary, so it is mainly used:

- When the cement that remains in the casing does not have to be drilled out.
- When an operation with a packer and a short tail pipe is thought to be too risky.

C. Slurry injection while circulating

This is in fact a special case that is related more to a circulation technique than to a squeeze operation. It can be used when cement needs to be restored over a considerable height (following faulty filling):

- either by using only lower perforations with returns at the top of the annulus, or
- by using lower and upper perforations isolated from each other by a cement retainer.

In the second instance a cement retainer and not a packer is mandatory because there is always the risk of slurry flowing back through the upper perforations into the annulus. This is true even if the volume of slurry is held strictly to the theoretical volume needed to fill up the space behind the casing. Therefore, with a packer there would be a good chance of getting stuck.

The procedure is carried out in two stages (Fig. 2.10):

- establish (and test) circulation (Fig. 2.10a)
- pump the slurry into place (Fig. 2.10b, c and d).

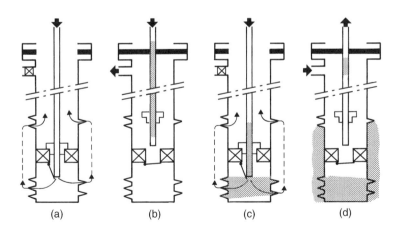

Fig. 2.10 Positioning the slurry by circulation (*Source:* ENSPM Formation Industrie).

D. Adding a bridge plug

A bridge plug is used when:

- One or more perforated (or weak) zones located under the treated zone need to be protected.
- The distance between the treated zone and the bottom of the well is too long.

When a retrievable bridge plug is used it is covered with sand before the slurry is pumped. In this way it is protected and the excess cement in the casing can be drilled out without damaging it.

2.2.2.4 Implementation (low-pressure squeeze)

A. Preparing for the operation

a. Gathering the data

As for any operation on a well, all the relevant data must first be gathered and a synthesis must be made. In particular, the exact objective of the operation and the conditions of the well need to be identified.

b. Slurry characteristics

A successful squeeze operation depends largely on the slurry's properties. The essential factors are:

- The compatibility between the slurry and its immediate environment in the well. Special attention must be paid to the quality of the mixing water and head and tail spacers need to be used to protect the slurry.
- The filtration rate (too little filtrate will give cement that is not properly dewatered, while too much may cause premature setting in the well before the slurry has had time to reach all the zones that are to be treated).
- The thickening time (it must be appropriate in relation to well conditions and the procedure that is used).
- The density and uniformity (cement quality depends mainly on the water/dry cement ratio).

The type of cement and the additives required can then be determined by laboratory testing. The products are virtually the same as for primary cementing.

c. Volume calculations

The volume of slurry needed is approximately 200 to 300 liters per meter of perforations (15 to 25 gal/ft), with a minimum of 1500 to 2000 liters (400 to 500 bbl). A larger volume is necessary if formation permeability is high or if the reservoir is naturally fissured. Additionally, the volume of slurry must be smaller than the amount that would cause hydrostatic pressure in the well to exceed the fracture pressure. When a squeeze operation is performed with the slurry still in the string, the volume needs to be smaller than the inside volume of the string. Also calculated are the volume to be circulated, the maximum volume to be squeezed, the displacement volumes, etc.

d. Pressure calculations

Depending on the volume of slurry pumped into the well, the maximum allowable wellhead pressure has to be determined such that the bottomhole pressure does not exceed the formation fracture pressure.

Other considerations are:

- the burst pressure of:
  - the wellhead
  - the string
  - the casing (depending on the procedure used)
- the collapse pressure of the casing (squeeze with packer or cement retainer).

Then the following points can be determined:

- the pressure, if any, to be applied to the annulus
- appropriate action in case of trouble and/or if the operation does not proceed according to the established program.

B. Preparing the well

a. Conditioning the well

After the casing has been cleaned and its diameter has been checked, the well is left full of a clean homogeneous fluid, if possible solid free. This is particularly important for the injectivity test.

b. Cleaning perforations

So that the slurry can flow through the perforations and come into contact with the formation, they must be cleaned before the squeeze. The methods used to clean perforations are covered in Section 2.3.4.3.

C. Injectivity test

This test is to be carried out:

- systematically
- with a fluid that is clean, solid free and compatible with the formation and the slurry
- before starting to mix the slurry.

The procedure is as follows:

- Run in, position and pressure test the squeeze string (also test the surface facilities and the annulus).

- Depending on the type of string, circulate the injection fluid as close as possible to the perforations (this can be very important depending on the type of fluid in place in the well).
- Mechanically isolate the perforated interval from the rest of the well (or depending on the equipment, close the BOPs).
- Try to determine:
  - the initial injection pressure (and compare it with the fracture pressure),
  - the injection flow rate that can be obtained for a stabilized bottomhole pressure lower than the fracture pressure.
- If the injectivity test is positive, return to the circulation position, if needed reverse circulate to get rid of the excess injection fluid and get ready for the squeeze (prepare the slurry, etc.).
- Otherwise clean the perforations and run another injectivity test.

D. *The squeeze operation itself*

Only a few particularly important operating points are repeated or mentioned here since the basic procedure was dealt with previously according to the string that is used:

- Record all parameters (pressures, flow rates, etc.).
- Prepare the slurry (measure its characteristics and take samples):
  - after getting a satisfactory injectivity test (pay attention to the mixing water too)
  - in a small tank
  - with good quality products.
- Use head and tail spacer fluids that are compatible with the fluid in place in the well and with the slurry.
- Monitor the pumped volumes closely:
  - when the slurry is positioned by circulation (when a packer and a short tail pipe or a cement retainer are used, the returns must absolutely be controlled by choking on the annulus)
  - during the squeeze.
- Reverse circulate to clean out the pipe as soon as possible (before or after squeeze depending on the equipment used).
- During the squeeze, monitor drillpipe and annulus pressures closely so that maximum allowable pressures are not exceeded, in particular fracture pressure.
- If need be, pressurize the annulus moderately during the squeeze.
- If possible, begin hesitation squeeze only after the slurry has come out of the string.
- When a packer plus a short tail pipe or a cement retainer are used and if the final squeeze pressure is reached before the volume of slurry has all been injected:
  - check that there are no returns through the drillpipes
  - reverse circulate immediately
  - depending on the case and the equipment, repressurize the well moderately for a few hours.

- If the final squeeze pressure is not reached before all the useful slurry volume has been injected:
  - anticipate the fact by starting hesitation squeeze (if this has not already been done)
  - rely on dewatering later on or, to be safe, overdisplace the slurry beyond the perforations, wait until the cement has set sufficiently and then repeat the squeeze with a prior injectivity test and redefine the slurry slug that is to be pumped (volume, etc.).
- Pull out the string while waiting for the cement to set.
- Run in a redrilling bit and a scraper if necessary, finish waiting for the cement to harden, if need be, and drill out the excess cement.
- Immediately draw up an accurate report of the operation, fill in the P and Q versus time diagrams.
- Check the quality of the restored cement job (see following section).

E. Final check

The effectiveness of the squeeze operation must be checked after:

- waiting for the cement to harden
- drilling out the excess cement (if this is scheduled).

The methods that can be used are the same as for initial evaluation.

2.3 PERFORATING

The aim of perforating is to re-establish the best possible connection (through the casing and the sheath of cement) between the pay zone and the borehole when the chosen configuration is cased hole.

Although perforating was done originally by bullets and even though in some very special cases other techniques such as hydraulic perforation may be advantageous, today shaped charges are used almost exclusively.

An effective connection depends largely on the choice of the perforation method and of the type of support or gun.

2.3.1 Shaped charges

2.3.1.1 Principle

The shaped charge is made up of five components (Fig. 2.11):

- the main explosive charge
- a cavity covered with a cone-shaped metal liner
- a primer charge (fuse)
- a detonating cord
- a case.

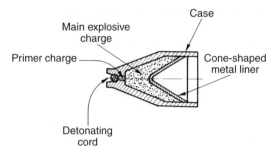

Fig. 2.11 A shaped charge (*Source:* After Schlumberger's "The essential of perforating").

When it is detonated, a high velocity (approximately 7000 m/s, or 20 000 ft/s) jet of gas shoots out of the tip along the axis of the cone due to the metal liner.

A perforation is the result of the jet impinging on the target material in front of the charge. It is produced by the jet's pressure (approximately 30 000 MPa or $5 \cdot 10^6$ psi) and depends on the length and width of the jet. Because of this, there will be some commotion in the vicinity of the perforation, the rock will be fractured and considerably compressed.

The jet of gas is followed by a slug (the melted residue of the metal cone) which travels much more slowly (approximately 300 to 1000 m/s, or 1000 to 3000 ft/s). The slug as such is of no importance, except that it may obstruct the end of the perforation.

Shaped charge performance mainly depends on:

- the amount of explosive load
- the type and angle of the metal cone (by varying it, perforation diameter can be prioritized over penetration depth and vice versa)
- the distance between the shaped charge and the target
- the density of the target.

For example, a charge the size of a hen's egg containing approximately 22 g (3/4 oz) of explosive can give a perforation about 1 cm (0.4 in) in diameter and 40 to 70 cm (16 to 28 in) penetration depth through the casing, cement sheath and into the formation. Performance is characterized according to the API RP 43 standard which is discussed in the next section (API: American Petroleum Institute, RP: Recommended Practice).

2.3.1.2 The API RP 43 standard

Section one of the standard covers tests run under surface conditions on a target consisting only of a casing and some concrete. The tests supply information only on the entrance hole diameter and the penetration depth versus the clearance of the charges.

Section two deals with tests run:

- on a sandstone target cemented in a steel canister
- in the presence of a fluid and under specified pressure and temperature conditions.

2. CONNECTING THE PAY ZONE AND THE BOREHOLE

Diagrams 2.1a and 2.1b illustrate the pressure conditions prevailing when the tests are carried out depending on the situation in which the charges will be used.

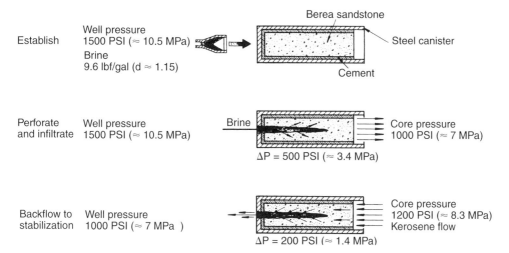

Diagram 2.1a Positive pressure flow test procedure.

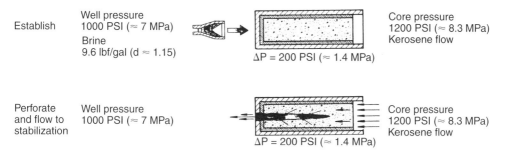

Diagram 2.1b Reverse pressure flow test procedure
(*Source:* After Schlumberger's "The essential of perforating").

These tests give a number of geometric characteristics (Fig. 2.12):
- the Entry Hole diameter, EH, usually from 0.25 to 0.5 in ($6 \cdot 10^{-3}$ to $12 \cdot 10^{-3}$ m)
- the total "useful" penetration into the core: TCP (Total Core Penetration), usually from 2 to 18 in (0.05 to 0.46 m)

- the total "useful" penetration into the target: TTP (Total Target Penetration)
- the "Overall" Penetration: OAP, that includes the final part of the perforation that contains the slug and the thickness of what represents the casing and the sheath of cement.

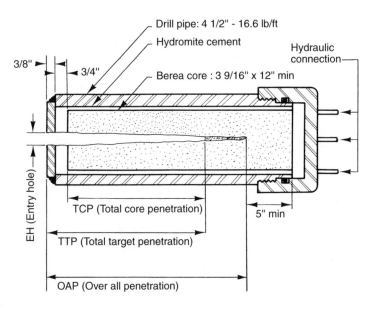

Fig. 2.12 Geometric characteristics of a perforation (*Source:* After Schlumberger's "The essential of perforating").

The tests also provide characteristics related to flow rate:

- the Core Flow Efficiency: CFE, usually 0.7 to 0.9, which corresponds to the ratio between the flow through the perforations after stabilization and the flow (under the same pressure conditions) through an "ideal" perforation (one that was drilled out) with the same geometric characteristics (i.e. the same EH and TCP)
- the "effective" penetration or equivalent penetration in the core: ECP (Effective Core Penetration) defined as follows: $ECP = TCP \times CFE$.

Care should however be taken is using these figures, especially the ones involving flow rate, since:

- The pressure and temperature conditions are not the same as the ones in the well.
- The core is made of a reference sandstone that has nothing to do with the reservoir that is going to be perforated (it may be limestone, etc.).
- The fluids used in the test do not correspond to the actual ones either.
- The CFE is measured for a ratio of stabilized flows with a specified pressure differential, but the ratio varies with the pressure differential and here again it most certainly does not correspond to the actual differential opposite the perforation when producing.

2.3.2 Main parameters affecting the productivity of the zone produced by perforating

2.3.2.1 Number of effective perforations

A fact that is unfortunately often overlooked is that the important thing is not the shot density but the number of effective perforations (in other words, that penetrate sufficiently into the formation and are not plugged up). Though there is a relationship with the shot density, it is mainly the conditions of firing and cleaning (see Section 2.3.4.3) that are determining. Wells have been known to have less than 1 to 10% of the perforations effectively flowing.

In most instances a density of 1 to 4 Shots Per Foot or SPF, (3 to 13 shots per meter) is enough. Without taking into account any type of damage, a density of 4 SPF (13 shots per meter) will in fact give a productivity index on the same lines as in an open hole.

A higher shot density may however be useful in reducing pressure losses in the perforations themselves and in their immediate vicinity, for example when:

- The volume flow rate per meter (per foot) of producing layer is very high (turbulence effect).
- It is necessary to use a sand control process (see Section 2.4.2).

2.3.2.2 Distribution of perforations over the producing zone (partial penetration effect)

This is also a fundamental parameter. For example, let us consider a stratum 100 m (or ft) thick that is perforated only over 20 m (or ft) of the total height. If the 20 m (or ft) that are perforated are spread out over eight zones 2.5 m (or ft) high separated from each other by a 10 m (or ft) non-perforated stretch, the productivity index will be two times higher than if the lower 20 m (or ft) alone were perforated. This is particularly significant if the vertical permeability is much lower than the horizontal permeability (usually true) or if the formation does not have a homogeneous vertical permeability (if there are shaly intercalations, etc.).

Although the perforation pattern is chosen according to reservoir considerations (interfaces and their variations, facies, etc.), it also depends in practice on the position and number of perforations that are effectively open.

2.3.2.3 Perforation penetration

If there is no damaged zone around the wellbore, this parameter is mainly significant for a penetration of less than 0.3 m (1 ft), but much less so beyond this figure. For example, going from 0.15 to 0.3 m (0.5 to 1 ft) penetration allows approximately a 20% increase in productivity index.

In fact what is of prime importance is to compare the penetration depth and the depth of pay zone damaged during drilling and casing operations, provided that the perforation itself has not been plugged.

Penetration is mainly dependent on:
- the explosive load
- the shape and type of cone
- the clearance between the cone and the casing and therefore (Fig. 2.13):
  - the size of the support or gun in relation to the casing
  - the number of shot directions.

It should be pointed out that the actual depth depends on the compressive strength of the rock and therefore does not correspond to the depth given by the API RP 43 test.

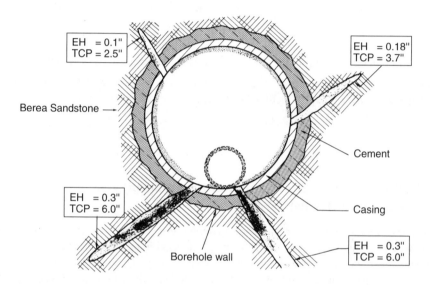

Fig. 2.13 Effect of the gun size and the number of shot directions on penetration (*Source:* After Schlumberger's "The essential of perforating").

2.3.2.4 Characteristics of the crushed zone

Because of the very nature of the shaped charge method, the immediate vicinity around the perforation is damaged. This is expressed to a certain extent by the CFE (see the reservations mentioned above). This type of damage depends on the charge itself (type of explosive and especially the shape and type of cone) and on the target formation.

2.3.2.5 Number of shot directions

Going from one to two shot directions (located at 180°) raises the productivity index by about 20%. Beyond this, (three [120°] or four [90°] shot directions), the increase is only slight.

In practice, the following are used because of penetration:
- one-directional guns if they have a small diameter in relation to the target casing
- multidirectional guns when large diameter guns (in relation to the target casing) can be used.

2.3.2.6 Perforation diameter

From a diameter of 6 mm (0.25 in) and above, this parameter usually has little impact. However, it is essential because of pressure losses:

- For sand control: here what are called big hole charges are used, giving an entry hole diameter of approximately 25 mm (1 in) at the expense of penetration.
- When the flow per perforation is high, but this corresponds to an extremely high flow (several tens of cubic meters or bbl of oil per day per perforation and several tens of thousands of standard cubic meters, of several hundreds of stcuft of gas per day per perforation).

The entry hole diameter is mainly related to the cone angle of the charge and the clearance between the cone and the casing.

2.3.3 Perforating methods and corresponding types of guns

The choice of the method is the result of a tradeoff between:

- well constraints:
  - reservoir pressure (normal or high pressure)
  - type of effluents: oil, gas, water, with or without H_2S
  - whether there was any previous plugging
  - reservoir thickness, porosity, permeability and homogeneity
  - situation of the casing and the cement job
  - risks of sand intrusion
  - whether the well is a producer or an injector
  - safety
- and optimum perforating conditions (which are not necessarily compatible with one another):
  - underbalanced shooting (1 to 2.5 MPa or 150 to 350 psi for an oil well, more for a gas well)
  - clean fluid in the well
  - large-diameter perforator loaded with high-performance charges
  - at least two shot directions
  - perforation clearing as soon as possible after shooting.

We will restrict our discussion to the three most conventional basic methods:

- positive or overbalanced pressure perforating before equipment installation
- reverse or underbalanced pressure perforating after equipment installation
- TCP perforating (Tubing Conveyed Perforator) run in on the end of the tubing.

2.3.3.1 Overbalanced pressure perforating before equipment installation

Perforations are made before running the downhole equipment (tubing, etc.) while the well is full of completion fluid exerting hydrostatic overpressure to counter reservoir pressure (Fig. 2.14).

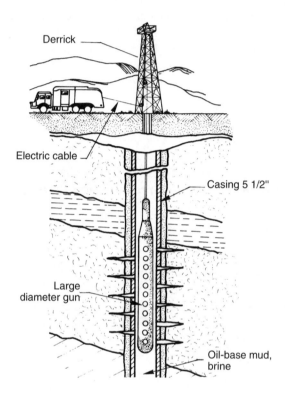

Fig. 2.14 Overbalanced pressure perforating before equipment installation (*Source:* After Schlumberger's "The essential of perforating").

Here a large gun can be run directly through the casing (casing gun) with the following advantages:
- large explosive loads
- multiple shot directions with close clearance and consequently good geometric characteristics, particularly penetration.

However, since perforating occurs under overbalance conditions, the perforations are plugged by the completion fluid present in the well (but since the pay zone has been drilled previously, it is easier to use a fluid that can cope with this problem). Thereafter it will be more difficult to get all the perforations to behave actively (see Section 2.3.4.3: Cleaning the perforations) and this may consume a lot of rig time.

2. CONNECTING THE PAY ZONE AND THE BOREHOLE

In addition, for further operation (running downhole equipment, replacing the BOPs by the production tree, etc.) safety conditions will not be as good since the borehole and the pay zone have already been connected.

Retrievable carriers are used (Fig. 2.15) that normally allow a shot density of up to 4 SPF (13 per meter) and even more, particularly for sand control (12 SPF), and shot angles of 90°, 120° or 180° depending.

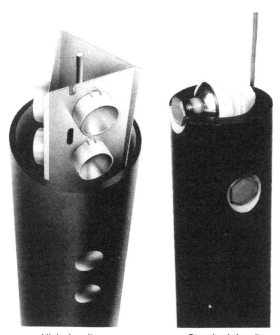

High density   Standard density

	Gun diameter (in)	Shot density (SPF)	Number of shot directions	Explosive load (g)	API RP 43 Section I			
					Casing size (in)	Casing weight (lbm/ft)	Entrance hole (in)	Penetration (in)
High Density	2 1/2	6	6	10.5	3 1/2	9.2	0.29	17.3
	2 7/8	6	6	15.0	4 1/2	11.6	0.30	22.0
	4 1/2	12	8	21.3	7	32.0	0.4	17.2
		12	6	24.0	7	32.0	0.7	7.9
	5	12	8	21.3	7	32.0	0.39	22.8
		12	8	24.0	7 5/8	33.7	0.61	9.8
	7	12	6	38.8	9 5/8	47.0	0.39	40.0

Fig. 2.15 Retrievable casing gun type carriers (*Source:* After a Schlumberger document).

These guns are leakproof, offer good reliability and leave no debris in the well. Their unit length ranges from 1.80 m to 3.30 m (6 to 11 ft) and they can be assembled. They are run in by electric cable and the different component guns can be fired selectively. However, the length that can be run in at one time is limited (approximately 10 m, 30 ft, or less, depending on the shot density, the well profile, etc.)

2.3.3.2 Underbalanced pressure perforating after equipment installation

Perforations are made after well equipment has been run in and once the production tree has been installed, with the well full of a "light" fluid (i.e. which exerts a hydrostatic pressure lower than the reservoir pressure). The guns are run into the well through the tubing (through tubing gun) by means of an electric cable through a lubricator (Fig. 2.16).

Because of the pressure underbalance, plugging is reduced or even avoided during and after firing. Additionally, since all the well equipment is in place there are no added safety problems and the well is ready to be cleared.

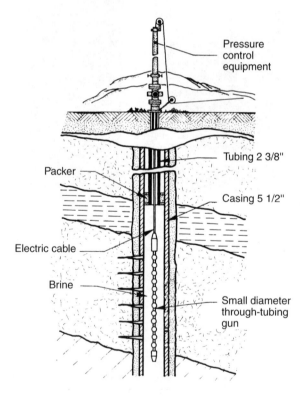

Fig. 2.16 Underbalanced pressure perforating after equipment installation (*Source:* After a Schlumberger document).

However, the guns are small in diameter compared to the casing since they must pass through the tubing. The charges are smaller and consequently the penetration is shallower and shots can be fired in only one direction because of the clearance.

Small diameter retrievable carriers can be used (Fig. 2.17a) but the charges are a major drawback as far as penetration is concerned. In addition, the guns have a tendency to expand when they are fired and retrieval through the tubing may prove to be problematical.

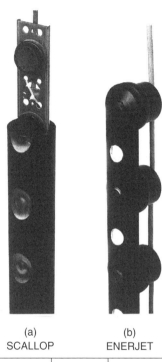

(a) SCALLOP (b) ENERJET

	Gun diameter (in)	Shot density (SPF)	Number of shot directions	Explosive load (g)	API RP 43 - Section I			
					Casing size (in)	Casing weight (lbm/ft)	Entrance hole (in)	Penetration (in)
Scallop	1 11/16	4	1	3.2	4 1/2	11.6	0.22	9.2
	2 1/8	4	1	6.5	4 1/2	11.6	0.25	13.5
		4	6	6.5	4 1/2	11.6	0.22	14.1
		4	6	6.5	4 1/2	11.6	0.36	5.9
Enerjet	1 11/16	4	1	8.0	4 1/2	11.6	0.26	16.7
	2 1/8	4	1	14.0	5 1/2	17.0	0.30	27.5
	2 1/2	4	8	21.0	7	32.0	0.36	28.4

Fig. 2.17 Carrier of the through tubing gun type (*Source:* After a Schlumberger document).

This is why semi or fully expendable carriers are used (Fig. 2.17b) that can carry larger loads, but:

- leave debris in the well
- cause more deformation in the casing and cement sheath
- do not protect the charges from the well's pressure and temperature, and as a result the charge efficiency is a little more restricted.

2.3.3.3 TCP perforating (Tubing Conveyed Perforator)

Here the guns are run in directly with the tubing (Fig. 2.18a). This method combines the advantages of the first two since:

- Large diameter guns can be run in.
- They can be fired with underbalanced pressure and with the permanent well equipment in place if so desired.

Very long stretches of casing can also be perforated in one single operation (several hundreds of meters and even more than one thousand meters in case of horizontal wells, i.e. one to over several thousands of feet), high shot density can be used, considerable pressure underbalance can be applied, the guns can be run in highly deviated wells, etc.

However, there are some drawbacks that are far from negligible:

- With permanent equipment in the well, if access is required opposite the pay zone for wireline jobs (production logs, reperforating, etc.) it is necessary to release the gun after shooting and let it fall to the bottom of the well (Fig. 2.18b and c). This will entail extra costs because a "trash dump" has to be drilled (an extension of the borehole below the producing stratum, which serves to collect all the objects and items such as sediments, etc. that are accidentally or deliberately left in the bottom of the well during completion or production, so that said objects and items cause the least possible trouble because they are located below the pay zone).
- Charges' temperature resistance and performance decrease with time. Here, since they are run on the tubing rather than with a cable, a lot more time is required to get them to the bottom of the well.
- It is impossible to check that all the charges have been fired except by pulling out the equipment.
- If the guns are not fired, pulling out the tubing is time-consuming and causes safety problems.

In practice the TCP guns are mainly used with a temporary string:

- To perforate before equipment installation when a long stretch of casing is involved or when large-diameter perforations are required with a high shot density (for sand control by gravel packing for example).

2. CONNECTING THE PAY ZONE AND THE BOREHOLE

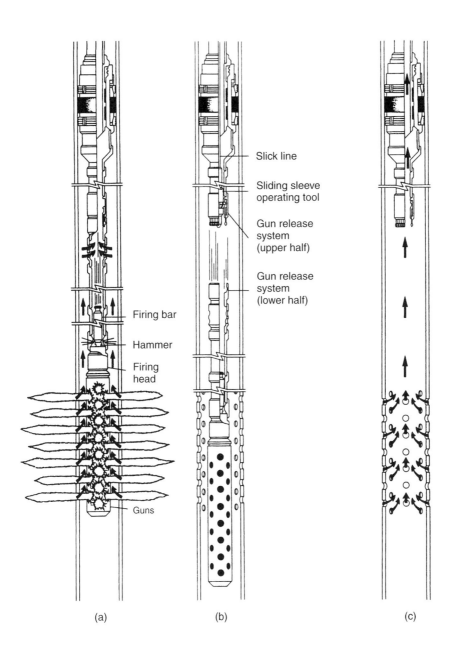

Fig. 2.18 TCP perforating (*Source:* After a Baker catalog). a. Firing. b. Releasing the gun. c. Producing.

- To perform a cased hole well test of the DST type (Drill Stem Test, a test carried out with test tools run on drillpipes) and perforations in one single operation, with a subsequent gain in safety and time, provided the recorders are not damaged by the shock waves from the explosion of the charges.

The TCP method uses specific equipment such as:

- guns (that look like the large diameter retrievable carriers seen earlier)
- firing heads
- a gun release system
- circulating devices, with or without a rupture disk (to circulate a light fluid in the tubing before the packer is set in order to obtain underbalanced pressure when firing: the rupture disk protects the mechanical firing head from any debris that might accumulate above it)
- tubing isolation devices (that allow the tubing to be run more or less empty, they are opened after the packer is set to put the well in underbalanced pressure conditions before firing)
- shock absorbers (to protect the equipment that is located above, particularly the well test recorders)
- a radioactive depth reference (to position the guns with a high degree of accuracy).

Depending on the case, the above equipment can be actuated:

- mechanically (by dropping a sinker bar, by wireline)
- hydraulically (by tubing or annular pressure)
- electrically (by dropping a sinker bar with an incorporated battery, by electric cable)
- automatically by the energy released when the shaped charges explode.

2.3.4 Specific points in the operating technique

Without going into the details of operating technique, we would like to touch on some particularly important points:

- safety
- adjusting perforation depth
- cleaning perforations
- monitoring the results.

2.3.4.1 Safety

Shaped charges are dangerous by nature and using them requires compliance with safety practices.

A. Electrical system check before perforating

The aim of checking up the electrical system is to prevent any risk of eddy currents (equipment is earthed, no welding allowed, no improvised electrical wiring, etc.).

B. Basic safety

Perforating operations are not performed:
- during storms (for the same reasons as previously mentioned)
- at night, except if the reservoir is already known and provided appropriate measures have been planned (emergency evacuation procedure, adequate lighting, etc.).

When the operation is carried out with overbalanced pressure before equipment installation, basic safety is ensured by:
- the completion fluid in the well
- the drilling BOPs
- a high-pressure pump connected to the well on stand-by.

In addition, the stability of the well is monitored each time the gun is fired and while pulling out of hole.

When the operation is performed with underbalanced pressure after equipment installation, basic safety is ensured by:
- the production wellhead along with a high-pressure pressure-tested lubricator with a stuffing box and a BOP adapted for use with an electric cable
- a high-pressure pump connected to the well on stand-by.

Additionally, the wellhead pressure is monitored each time the gun is fired and while pulling out of hole.

C. Further precautions when loading, starting to run in and concluding pull out

During these phases further measures are also necessary:
- Cut off all radio broadcasting (notify the base and logistics: helicopter, boats, etc.).
- Get non-essential personnel out of the way.
- Keep anyone from standing in the line of fire.
- Take extra care while pulling out if there has been a misfire.

2.3.4.2 Perforation depth adjustment

Perforations are not positioned by directly measuring the depth from the surface, but by referring to the logs (electrical measurements) that were run for the reservoir department. All the logs (some before casing was run in, some afterwards) that allow the reservoir department to determine which reservoir zones are to be produced are matched to a reference log. This is usually the gamma ray (the γR measures the natural radioactivity of formations), which can be recorded either in an open or cased hole. Once the hole has been cased a CCL (casing collar locator, an electromagnetic tool that detects casing connections) is added to the γR. The perforator is then run along with a CCL alone or a CCL-γR. It is therefore accurately positioned with respect to the zone that is to be perforated.

2.3.4.3 Cleaning the perforations

The state of the perforations after firing mainly depends on the method used for perforating (over- or underbalanced pressure) and on the type of fluid in place in the well.

Laboratory studies have shown that:
- When perforating underbalanced in clean brine, the perforations can be made to flow as soon as a pressure differential is exerted. Moreover, there is only a small reduction in the productivity index.
- In contrast, when perforating overbalanced, depending on the type of fluid in the well and the exposure time, the required reverse pressure may be several MPa (several hundred psi) and the productivity index may be considerably reduced (by up to over 60%). The figures are even more unfavorable if the produced fluid is a gas instead of an oil or if the formation permeability is middling or low (less than 100 mD).

When cleaning is required after shooting, one of the methods listed below is usually implemented.

A. Well clearing

The well is made to flow through a large diameter choke so that the perforations are exposed to maximum reverse pressure. However, as soon as a few perforations start flowing they limit the reverse pressure that can be applied to the others. It is more difficult to lower the bottomhole pressure and, as it actually becomes lower, the pressure in the reservoir near the wellbore also tends to decrease. In addition, there would be a risk of detaching fine particles from the matrix, not being able to get them out of the formation and having them plug up interpore connections — especially in insufficiently consolidated formations.

B. Back surging

This technique consists in using a temporary string equipped with an atmospheric pressure chamber on the lower end (Fig. 2.19). By opening the lower valve, a considerable negative pressure is applied almost immediately to all of the perforations while the flow volume is restricted at the same time.

Another variation consists in running in a tubing that can be filled to a certain height via a bottomhole valve or a rupture disk. The valve or disk is opened suddenly once the equipment is in place.

Note that the pressure differential that is usually recommended in the literature is quite considerable (from 1.5 to 3.5 MPa or 200 to 500 psi for an oil reservoir whose permeability

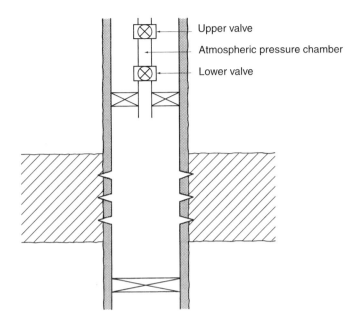

Fig. 2.19 Cleaning perforations by back surging (*Source:* ENSPM Formation Industrie).

is higher than 100 mD, from 7 to 14 MPa (1000 to 2000 psi) if its permeability is less than 100 mD; while for a gas reservoir the figures are respectively 7 to 14 MPa or 1000 to 2000 psi and 14 to 35 MPa or 2000 to 5000 psi). There is therefore a risk of destabilizing the matrix.

C. Circulation washing with a washing tool

The perforations are cleaned by circulating from one to the other beginning at the bottom by means of a tool equipped with cups (Fig. 2.20). The circulating flow rate is in the range of several hundred liters per minute (1 bpm). This technique is mainly used when gravel packing is due to be installed later on for sand control. It is designed to make sure that all the perforations are open, but the circulated fluid and the fines that are stirred up may damage the formation. Additionally, it requires a long time (besides the return trip, about 5 to 10 minutes per foot of perforations).

D. Acid washing

Acid is injected under pressure to restore the connection between the formation and the wellbore. Acid is pumped down to bottom either before final equipment installation by using a temporary string or after it (see Section 2.4.3.2 E).

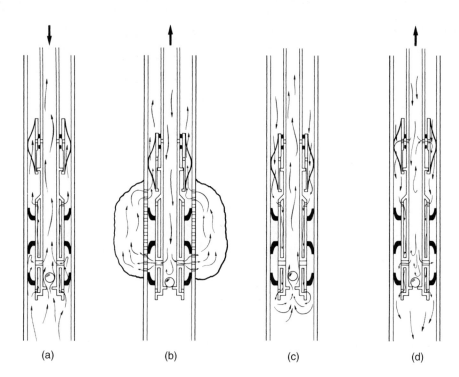

Fig. 2.20 Cleaning perforations with a washing tool (*Source:* Baker catalog). a. Running in. b. Washing perforations. c. Reversing off bottom. d. Coming out of hole.

In the second case, it is usually impossible to circulate (because of the presence of a packer and since a circulating device is not recommended for this type of operation) or to squeeze from the surface (it would be necessary to squeeze the whole tubing volume of completion fluid before the acid got to the perforations). A coiled tubing unit or, failing this, a snubbing unit must therefore often be used (see Chapter 5, Section 5.3).

Furthermore, acid naturally tends to penetrate the less damaged perforations and "forgets" the others (unless acid is injected between two squeeze packers that are moved along the height of the perforations, using a washing tool for example, or unless diverting agents such as ball sealers are used, etc.).

E. Conclusion

Whatever the cleaning method used, the perforations are never 100% unplugged and the modifications brought about by treatment may even promote plugging. Additionally, each time that cleaning is undertaken with a temporary string, temporary plugging agents may have to be used afterwards to restabilize the well so that the temporary string can be pulled out and the final completion equipment can be run in.

As a result, when the formation itself is not too badly plugged up, the best way of "cleaning" the perforations is to perforate under underbalanced pressure conditions after equipment installation to keep from damaging the perforations.

2.3.4.4 Monitoring the results

No specific type of monitoring is usually done. When completion is over or, sometimes, when under way, a well test is carried out. This helps assess the skin effect, giving an overall picture of perturbations of the flow in the vicinity of the wellbore due to several factors such as the formation plugging itself and partial penetration. Sometimes monitoring is restricted to merely gaging the flow.

Well testing and/or production logging (records of bottomhole flow rate, temperature, noise, etc. versus depth) zone by zone would be required in order to better evaluate the effectiveness of each perforation. However, logs are costly and also have their limits.

2.4 TREATING THE PAY ZONE

2.4.1 Problems encountered

The main purpose of treating pay zone is to try to solve two problems that may even be combined:

- insufficiently consolidated formation
- insufficient productivity.

They are due to original intrinsic factors or factors induced during drilling, completion or production. Several basic phenomena can be involved.

2.4.1.1 Phenomena pertaining to insufficient consolidation

A. Initially insufficient consolidation

The rock may be insufficiently consolidated naturally, depending on:

- the type and amount of intergranular components
- the burial depth and therefore the pressure and the temperature that the formation has been subjected to.

B. Modification in intergrain bond stresses

This may be due in particular to:

- the flow rate and viscosity of the moving fluid
- the partial transfer of some of the overburden from being supported by the pressure to being supported by the matrix during depletion.

C. Drop in mechanical properties

"Secondary" fluids (water, water with different salinity, etc.) have a dispersing or sloughing action on some mineral components (shales, feldspars, etc.) that help bind the grains to each other. The result is a weakening of the formation.

2.4.1.2 Phenomena pertaining to insufficient productivity

A. Low absolute permeability (initial or caused by some disturbance)

Absolute permeability is a rock's permeability when it is entirely saturated with a single one-phase fluid. It can be low either naturally or due to some damage. Fluids moving in the pores come up against very severe conditions such as twisted interpore passageways, rough pore walls and various minerals that can react.

A drop in absolute permeability may therefore be due to fine solid particles getting trapped in the pores. These solids may come from the fluids used in the well during the various operations that have been carried out (filtration, losses, injections, etc.). They may also come from the reservoir itself. In addition to swelling and dispersion of some shales in contact with "fresh" water, there are simple changes in salinity or flow rate which can help release particles that are loosely attached to the walls. These particles will restrict the flow where pores narrow down.

Contrary to common practices, in some cases well clearing at a low flow rate gives better results than at a high rate.

A drop in absolute permeability can also be due to precipitates if the salts in solution in the fluids in place are incompatible with those in the fluids that penetrate the formation during the various operations carried out in the well (including treatment).

It should be pointed out that permeability is not the same vertically and horizontally or in all directions.

B. Low relative permeability (initial or caused by some disturbance)

When several fluids are present, the permeability to one of them is called effective permeability (it is lower than absolute permeability and depends on the fluid's saturation). The relative permeability is the ratio between the effective and the absolute permeability.

Relative permeability is affected by:
- The respective saturation in the different fluids that are present, for example increased saturation in water (due to a filtrate, injection, etc.) means a decreased permeability to oil or to gas and an increased permeability to water (Fig. 2.21).
- Wettability of the walls, which means the walls' ability to be covered by a film of oil rather than a film of water and vice versa. Oil or gas wettability lowers the corresponding relative (oil or gas) permeability. Wettability can be influenced by the presence of surfactants.

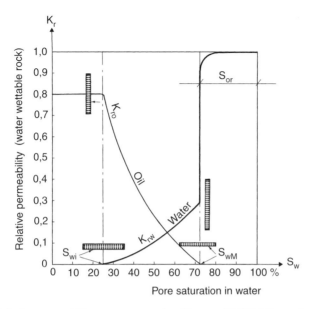

Fig. 2.21 Relative permeability versus saturation (*Source:* ENSPM Formation Industrie).

C. High viscosity (initially or due to some disturbance)

Viscosity depends on the hydrocarbon's composition (gas, oil, heavy oil) and on temperature. When an emulsion is formed with water, viscosity can increase considerably. Here again the presence of surfactants is very important. Depending on the case, surfactants can promote the emulsion's formation and stability or prevent it.

D. Effect of perforations

Here there are mainly two factors:

- Damage to the perforation itself (crushed and compacted zone with lower permeability around the cavity created by the explosive charge, dewatered mud in the cavity, etc.), along with the number of perforations actually opened per unit of casing height.
- The partial penetration effect when only part of the pay zone thickness has been perforated.

2.4.1.3 Main causes of problems

These problems can have a number of causes, the main ones are discussed below according to when they may crop up.

A. Existing initially

- little or no consolidation
- low natural permeability
- mineralogical composition, in particular presence of mobile or loosely attached particles
- oil wettable formation
- presence of viscous fluids or of components that promote emulsions
- salts (that can precipitate out or flocculate) in the reservoir.

B. Arising during drilling

- solid particles in the drilling mud
- filtrate reacting with the matrix or the fluids in place
- contact time
- overpressure on the stratum
- fine particles due to the formation (and mud solids) being ground down
- use of a lost-circulation material (LCM) to reduce losses
- cementing.

C. Arising during completion

- perforations
- cement squeeze job
- completion fluids (composition, cleanness)
- use of temporary lost-circulation material
- acidizing side reaction
- implementation of sand control
- cleanness of equipment
- running in and pulling out.

D. Arising during production

- decreased reservoir pressure
- water production
- various deposits (carbonates, salts, paraffin, etc.)
- gradual plugging of the sand control device
- corrosion
- bacteria
- injection water (type, cleanness).

2.4.1.4 Means of diagnosis

A number of different means are used to diagnose the quality of the pay zone-borehole connection in order to determine the problems that may arise as well as their causes. It is crucial to do this to define what preventive measures should be implemented (in particular regarding the quality and cleanness of fluids and equipment), and if need be how to choose appropriate remedial action.

A. Study the core samples

Core samples taken when the pay zone is being drilled give some information on the reservoir, in particular on its mineralogical, petrophysical and mechanical characteristics.

B. Study the fluids

Provided representative samples have been taken, PVT studies (Pressure, Volume, Temperature) and other laboratory analyses (reservoir water analysis, emulsion test, etc.) allow the type of fluids in place and their behavior (alone or in conjunction with others) to be identified.

C. Well testing

Besides giving the reservoir pressure figure, well testing also helps distinguish between low natural permeability and problems in the vicinity of the wellbore when productivity is insufficient. However, the kind of problems still remains to be identified (plugging in the formation itself, perforation efficiency, partial penetration effect, etc.).

D. Well history and production records

These are a very important source of information. An attempt must be made to find out exactly what happened during drilling, completion and production. The critical importance of accurate and thorough operations reports can not be overemphasized.

Monitoring the variations in the water cut and the solids content in the effluent, especially in poorly consolidated formations, can help detect potential problems or check that the amount produced is staying within acceptable limits according to the experience gained on the field.

2.4.2 Main types of remedial action for poor consolidation: sand control

2.4.2.1 Why control sand?

In poorly consolidated sandstone formations, grains of sand that make up the sandstone are pulled loose from the formation and are carried off by the fluid flow. The sand causes a lot of problems that can mean a considerable decrease in productivity and adverse safety conditions. Costly workover operations may become necessary as a result. Some of the problems that sand causes are listed below:

- Caves are formed (the borehole gets significantly wider because the walls slough off) and surrounding formations collapse, thereby decreasing productivity (surrounding shales damaging the connection between the borehole and the pay zone). The reservoir's cap rock may even start leaking (in the case where it is relatively thin and collapses).
- Sand accumulates in the bottom of the hole, thereby decreasing the effective height of the well opposite the pay zone.
- Sand accumulates in the tubing, plugging the well (particularly during temporary production shut down).
- Sand erodes the equipment, especially in the wellhead, in elbows and in the tubing, with the attendant blowout risks.
- Corrosion is accelerated.
- Higher risks when downhole measurements are made, or they may even become impossible.
- Higher risks when fishing or workover operations are carried out.
- Deposits are laid down in the flowlines or in the process facilities equipment.

Sand production must be monitored and kept within acceptable limits so that the well can be produced rationally.

2.4.2.2 Sand control processes

The basic processes can be broken down into two main categories:
- Mechanical, that consist in installing a filter (by far the most common).
- Chemical and physicochemical, that consist in reinforcing or stabilizing the intergrain bond.

Whether the completion is of the open hole or cased type influences the implementation and effectiveness of the process, and therefore the choice of the process itself.

A. *Processes for open hole completion*

a. Screens alone (Fig. 2.22a)

Figure 2.23 illustrates some types of screen. To control formation sand effectively and limit filter plugging, the slot of the screen will usually be chosen to retain the coarsest 10% approximately of the sand, which in turn controls increasingly finer grained sand.

Although this technique gives good results for coarse sand with a relatively homogeneous grain size, it does cause productivity to fall when the sand is fine grained or when there are shaly intercalations (because the screen or the sand pack around the screen gets plugged).

b. Gravel packing (Fig. 2.22b)

Gravel packing is placed against the formation and is held in place by screens. The gravel acts as a filter and the screen serves only to keep it in place. The gravel must be clean, calibrated with a selected grain size in relation to the formation sand.

2. CONNECTING THE PAY ZONE AND THE BOREHOLE

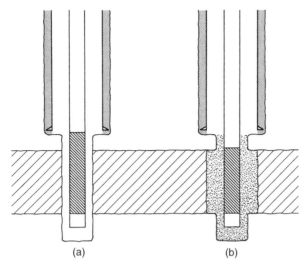

Fig. 2.22 Processes that can be used to control sand in open hole completion (*Source:* ENSPM Formation Industrie).
a. Screens. b. Gravel packing.

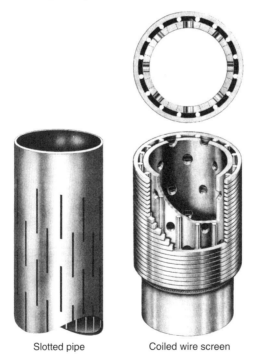

Slotted pipe   Coiled wire screen

Fig. 2.23 Screens (*Source:* Howard Smith catalog).

To avoid confusion in terminology, we use the term sand for the material that makes up the formation and the term gravel for the material that is positioned from the surface. However, the mean diameter of the "gravel" is approximately only a millimeter (0.04 in), the general sizing rule being that the gravel should be about six times coarser than the formation sand on the average.

This method allows the use of screens with larger slots than when used alone, thereby offering a larger cross-section to fluid flow. Permeability will also not be affected as much as it is with the first method. As a matter of fact, when screens are used without packing, it is mainly the fine particles that are carried off by the fluid and fill up the space between the formation and the screen.

The borehole is usually reamed out before the gravel is positioned so that there is enough space for a satisfactory thickness of gravel.

B. Processes for cased hole completion

A pay zone-borehole connection consisting of a perforated casing can not be considered as a sand control process. It can of course be hoped that arches will be formed opposite the perforations and thereby stop sand intrusions (especially if a large number of small diameter perforations have been made). However, these arches are in fact highly unstable, they are particularly sensitive to variations in flow rate and changes in water cut. This is why other techniques are needed for long-lasting sand control.

a. Screens alone (Fig. 2.24a)

The critical point is when the fluid passes the restricted cross-section of the perforations, which will be mostly obstructed by the produced sand. The sand that is retained by the screen slots fills up the space between the screen and the casing little by little and then fills up the perforations.

Pressure losses introduced at this point are usually prohibitive for liquids (oil, water) and so the process is normally not suitable in this case. In contrast, since gas has a very low viscosity the process may fill the bill in gas wells, particularly when the reservoir is perforated over considerable height and if the flow per unit of height is low.

b. Gravel packing (Fig. 2.24b)

As for screens alone, perforations are the critical point so the gravel emplacement technique must allow the perforations to be effectively filled up with gravel while at the same time maintaining its high permeability. For example a 20-40 mesh gravel (i.e. grain diameters ranging from 0.84 to 0.42 mm, or 0.033 to 0.017 in), which is a conventional grain size, has a permeability of about 150 darcys under surface conditions. This should be compared with the few tens or hundreds of millidarcys that is typical of sandstone permeability.

The cross-sectional area must also be as large as possible for the fluid to pass through, so a high shot density is recommended (conventionally up to 12 SPF, i.e. 39 shots per meter, is used), along with big hole type charges that prioritize entry diameter at the expense of penetration depth. Here, the TCP method (Tubing Conveyed Perforator) is particularly well suited. Another very important point is to unplug the perforations properly before installing sand control (using a washing tool, etc.).

c. *Consolidation* (Fig. 2.24c)

The aim of this method is not to create a filter to retain the produced sand but rather to reinforce or stabilize the bond between the grains of sand that make up the sandstone formation so that they are not entrained by the fluid flow. A thermosetting resin is often injected into the formation. Products can also be injected that tend to stabilize the components of the intergrain cement. This is achieved mainly by lowering their sensitivity to water.

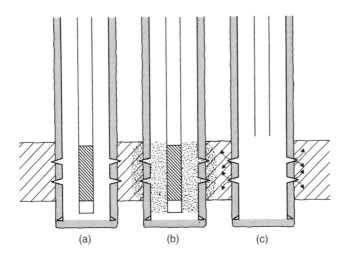

Fig. 2.24 Processes that can be used to control sand in a cased hole completion (*Source:* ENSPM Formation Industrie).
a. Screens alone. b. Gravel packing. c. Consolidation.

The main difficulty with these processes is to get the treatment product injected into all the perforations. As for any injection treatment, the injected fluid tends to penetrate only the most permeable zones and bypass the others. However, during the production phase, if just one untreated perforation produces sand it is enough to cause the process to eventually fail! Consequently, these processes are hardly ever used except on cased, perforated, homogeneous zones. They must also extend over only a short height so that there is a small surface to treat and a good chance of treating all of it.

Another difficulty in using resins is to inject enough over the total height of the treated zone to consolidate the formation satisfactorily but not enough to reduce permeability significantly. The resin injection method uses:
- Either a resin dispersed in a carrier fluid (solvent): after separation the resin coats the solid particles and gets concentrated at the contact points between the grains.
- Or a high-density resin, with the excess displaced farther inside the formation by pumping a neutral fluid (such as diesel oil, etc.) before the resin sets.

For the treatment to be effective, the reservoir must first be unplugged (and recompacted if it has already produced sand). Depending on the type of resin used, the in-place water must also be eliminated. In addition, a number of additives may be needed, such as catalysts, accelerating agents or bonding agents (resin-grain bonding).

d. Coated gravel process

Let us mention another process using resins. On the surface a mixture of resin and gravel is prepared. It is pumped into the well then forced onto the formation under pressure after prior preparation (conventionally, the perforations are washed with a washing tool). After the resin has set, the excess in the well, in particular opposite the perforations, is drilled out.

C. Process choice

From the standpoint of specific performance (effective sand control, productivity index), the best choice in most cases is gravel packing in an open hole.

However, in many cases it is necessary to case the pay zone (because of interfaces, heterogeneous reservoirs, multiple pay zones, etc.). Cased hole gravel packing is then usually chosen, or sometimes gravel packing with a window in the casing.

Attention is required, however, when boreholes are deviated more than 30°, problems arise in setting the gravel pack.

Screens alone may sometimes be suitable, mainly in an open hole or in a cased hole over a considerable perforated height, particularly when:
- Sand problems are not very severe (the formation has not yet produced sand, or it is clean and homogeneous, or the sand is coarse and homogeneous, etc.).
- The produced fluids have very low viscosity (gases) or low viscosity with low velocity (oil with a viscosity of around one centipoise produced at a velocity of a few centimeters or inches per second).

The other processes (consolidation, coating) are used only for certain special cases. Generally speaking, they are costly and implementation is tricky. They are suitable for coarse sands and a small casing diameter, provided there is not too great a height to be treated.

2.4.2.3 Gravel packing in a perforated cased hole

A. Selecting the gravel

The choice is based on the analysis of the formation sand, but getting representative samples is the problem! In this connection, preference is given to core samples and sufficient sampling density is required (each sample must be analyzed separately, do not use a composite sample). In some cases (particularly for homogeneous formations), sidewall core samples are acceptable. However, sand samples taken on the surface should never be used.

The basic analysis is sieve analysis, with the sample being sent through a series of sieves with increasingly fine mesh sizes. A grain size distribution curve is drawn on semilog paper representing the cumulative weight percentage of sand retained versus the mesh size (note that there are different units of measurement for sieve mesh size, with the most common being US mesh).

Other laboratory analyses are also carried out on intergrain cement, acid solubility, formation sensitivity to different fluids, etc.

A number of "rules" recommend choosing the gravel about five or six times bigger on the average than the formation sand and with a uniformity coefficient of approximately 1.5 (ratio between the mesh size that retains 40% of the gravel and the size that retains 90%). In fact the most "accurate" rules are based on a specific point of the grain size distribution curve for the formation sand (the 10%, 40%, 50% or 70%, etc. point) which depends on the uniformity coefficient of the sand. Besides size and uniformity, the chosen gravel must also have other characteristics such as grain shape, compressive strength, solubility in acids, etc.

B. Selecting screens

The screen slot, expressed in thousandths of an inch, which is usually chosen to correspond to two-thirds of the diameter of the finest part of the gravel selected (when screens are used alone, a screen slot size is often chosen equal to one or two times the 10% point of the formation sand).

Screen diameter is chosen to give the gravel a space of at least two centimeters (0.78 in) between the screen and the casing (or at least six centimeters, or 2.4 in, between the screen and the open hole, which usually implies reaming out). When screens are used alone, the outside diameter is chosen as close as possible to the hole diameter. The screens must be centered to have a uniform space all around them.

The screen length corresponds to the perforated height of casing plus one or two meters on either side.

The most commonly used screens today have a coiled wire, which provides a large inlet area and allows very slim slots. However, slotted pipe screens are much more economical and can be appropriate for coarse enough formation sand. It should be noted that the coiled wire screen diameter indicated usually corresponds to the carrier tube and not to the outside diameter of the screen.

C. Operating procedure

A number of points are essential for a gravel packing operation to be a success, particularly:

- Perforations (or formations) must be thoroughly unplugged before gravel packing as such is undertaken.
- The carrier fluid must be very clean (filtered to a few microns).
- The fluid circuit (tank, pump, gravel packing string, etc.) must be perfectly clean (acid washing if necessary).

Additionally, the displacement procedure must attempt to:

- fill perforations properly with gravel
- prevent the gravel from mixing with the formation sand
- prevent the gravel from bridging, as it would leave "empty" spaces.

The high density method is therefore preferably used, whereby the gravel is conveyed by a highly gelled fluid which allows high concentrations of sand to be kept in suspension (up to 15 pounds of gravel per gallon of fluid, i.e. 1.8 kg/l). The concentration is expressed in terms of weight of gravel to volume of liquid and not volume of mixture.

The gravel is conveyed to the bottom of the well by circulation at a moderate flow rate of 2 to 3 bpm (barrels per minute), i.e. 300 to 450 l/min approximately, then squeezed into its final position. Figure 2.25 illustrates this type of operating procedure and the relevant gravel packing string.

The low density method (concentration of 0.5 to 2 pounds per gallon, i.e. 0.06 to 0.24 kg/l) is used in some cases, particularly in an open hole or when the volume of gelled fluid would be too large. The sand is then positioned at a higher flow rate (to prevent sedimentation, since the carrier fluid is not gelled). The risks of mixing with formation sand, bridging and even faulty perforation filling are much higher.

2.4.3 Main types of remedial action for insufficient productivity: well stimulation

2.4.3.1 Principal types of well stimulation

The aim of these different methods is to improve a productivity or injectivity index that is deemed inadequate. They are designed to increase the recovery rate of reserves but not the recoverable reserves themselves, although they can make production of a given well profitable when it would otherwise have an insufficient flow rate. Before undertaking a stimulation treatment, it is essential to identify the problem properly so as to choose the type of treatment that can actually solve it.

Furthermore, it should be pointed out that a well's production depends of course on the productivity index, but also on the average reservoir pressure and bottom hole back pressure. As a result, a pressure maintenance operation, a change in tubing diameter, the implementation of an artificial lift process, etc. may have to be considered in addition to (or instead of) stimulation.

2. CONNECTING THE PAY ZONE AND THE BOREHOLE

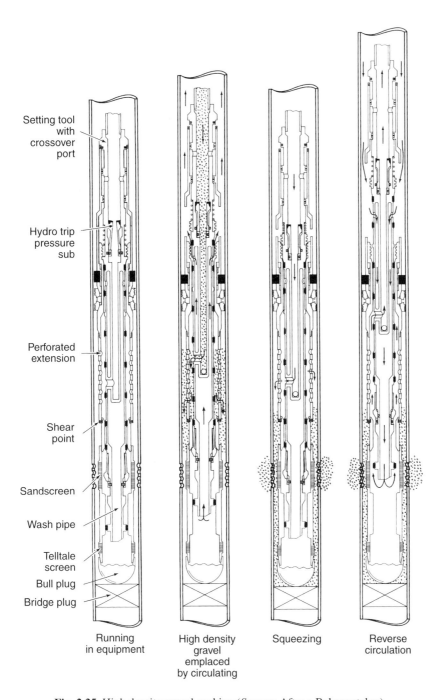

Fig. 2.25 High density gravel packing (*Source:* After a Baker catalog).

A. The influence of near-wellbore permeability variations on productivity

Reduced permeability in the near wellbore, even with a thin damaged zone, causes a considerable drop in productivity. In contrast, if the permeability is better (or even much better) around the wellbore than in the reservoir, the increase in productivity is limited.

Furthermore, although it is very easy to damage a formation's natural permeability seriously in the course of operations on the well (there are a large number of causes), on the contrary improving the natural permeability to any significant extent is much more difficult.

B. Matrix treatment

These are treatments (acidizing, solvent injection, etc.) carried out by injecting the relevant fluid at a pressure lower than the formation breakdown pressure. They act only in the vicinity of the wellbore and are effective mainly when formation plugging has occurred. If well suited to the problem, they will in fact allow productivity to be restored.

The better the formation's natural permeability, the greater the expected increase in productivity (if the operation is successful and in the case of the same reduced permeability around the wellbore).

The characteristics of the treatment fluid and the operating procedure basically depend on the cause and extent of the damage as well as on its location. The last point is important since the treatment fluid must remain effective while it is en route to the area due to be treated and because it must be able to act as efficiently as it can all throughout the area.

C. Deep treatments

These treatments (mainly hydraulic fracturing) are designed for low-permeability formations whose natural productivity is to be enhanced rather than restored. They consist in extending the well radius substantially by means of a more permeable drainway in the formation itself.

D. Other treatments

Inadequate productivity may be caused by other factors:
- When the problem is located at the perforations, washing them, reperforating (over the same height), or adding extra perforations (increasing the perforated height) may lead to a satisfactory result.
- When fluids are viscous or there are problems of interfacial tension, chemical injections, steam injection or in situ combustion can be envisaged.

2.4.3.2 Acidizing

A. Principle and application

Acidizing is a matrix treatment, i.e. a treatment carried out at a pressure lower than the formation breakdown pressure, whereby acid is injected into the formation to improve well productivity or injectivity. Actually, the process is mainly used when the objective is to restore near-wellbore permeability after formation plugging during drilling, completion or production.

In this case:

- Be sure that the reduced production is really due to formation damage and not some other factor (e.g. the well is sanding up, the water cut has risen, the reservoir is depleted or there is interference with a nearby well).
- Depending on the cause of plugging (e.g. paraffin deposited around the wellbore of a producing well), it may be necessary to use a solvent instead of acid.

Care should be taken as damage can be more severe after acidizing than before:

- Particularly in sandstone formations, because of the formation's deconsolidation, the release of fine particles, or side reactions that may produce precipitates, etc.
- Additionally, whether the formation is sandstone or carbonate, because of emulsion, water block (formation pores blocked by filtrate water, etc.), wettability inversion or impurities carried along into the formation (said impurities may come from treatment fluid that is not clean enough, from the pumping system, from the tubing or from corrosion of the equipment).

As a result, an acidizing treatment must not be decided on the spur of the moment, it must be studied and carried out carefully.

B. Types of acids used

The type of acid used depends on the type of formation rock.

a. Carbonates

From the standpoint of acidizing, rocks with over 20% of their components soluble in 15% hydrochloric acid are usually considered to be carbonates. Here 15% or even 28% hydrochloric acid is used. The concentration is adjusted on the site depending on the concentration of the commercial acid (i.e. about 35%), by adding the acid to the water and not the other way around.

Weak acids (5% hydrochloric acid, acetic acid or formic acid), mixtures of these acids, acid-based emulsions or acid gels are also used, particularly in hot formations, to slow down the reaction rate and thereby enhance treatment penetration depth.

These acids attack the carbonate formation directly and rapidly, but do not however react with the plugging solids, which are circumvented to restore proper connection between the borehole and the formation.

It should be noted that hydrofluoric acid should not be used in carbonate formations as it causes insoluble calcium fluoride to form.

There is no accurate rule to determine the amount of acid that should be injected. Generally speaking, 0.5 to 2 m^3 (40 to 160 gal) of acid solution is used per meter (per foot) of formation thickness to be treated. A cubic meter (gallon) of 15% hydrochloric acid can dissolve around 200 kg (1.7 lb) of carbonate rock.

b. Sandstones

From the standpoint of acidizing, rocks mainly made up of quartz, clays and feldspars, whose solubility in 15% hydrochloric acid is lower than 20% are usually considered to be sandstones. Here the most common solution is hydrofluoric acid based acids. On the site, hydrofluoric acid is prepared by dissolving ammonium difluoride in an aqueous solution of hydrochloric acid.

In practice, hydrofluoric acid mainly dissolves clays and feldspars (a cubic meter of 4% hydrofluoric acid can dissolve about 30 to 60 kg of clays, i.e. a gallon will dissolve 0.25 to 0.5 lb), since the solubilization rate of silica is too low at reservoir temperatures for it to be of any interest. However, in the presence of carbonates (and do no forget that carbonates form the most common cements in sandstones, along with silica!), hydrofluoric acid reacts very quickly to give a calcium fluoride precipitate. To prevent this, the hydrofluoric acid treatment phase is preceded by a hydrochloric acid preflush.

Other precipitates can also be formed, such as potassium and sodium fluosilicates (when the hydrofluoric acid concentration is high), precipitates of colloidal silica (in the presence of aluminum when the hydrofluoric acid concentration decreases), or precipitates of ferric hydroxide (when the pH becomes less acid), etc.

Additionally, the attack on intergrain cement, either during the hydrochloric acid preflush or during the main hydrofluoric acid treatment, may cause fine particles to be released and mechanical strength to be lost (sand influxes). This may result in a dramatic reduction in permeability.

Consequently, an acidizing operation on a sandstone formation should only be decided after careful study and laboratory tests. In particular, it is less the solubility rate than the Acid Response Curve (ARC) that is important. The ARC is a dynamic study that is used to determine the influence of acid drainage on a sandstone's permeability and consolidation.

In practical terms, a sandstone acidizing treatment involves three steps that are carried out directly one after the other without stopping:

- A preflush usually with 5 to 15% hydrochloric acid, designed to eliminate carbonates and prevent contact between the hydrofluoric acid and formation water (where sodium, potassium and calcium, etc. ions are present). The volume is equal to approximately half the volume of the main treatment.

- The main treatment designed to get rid of the damage. Depending on the formation's composition, 0.5 to 1.5 m^3 (40 to 120 gal) of mud acid, a mixture of hydrochloric acid (12% normally) and hydrofluoric acid (3%, sometimes 6%, even 10 to 15%), is generally used per meter (per foot) of treated formation thickness.
- An overflush or afterflush to maintain an acid environment around the well and/or displace the side reactions related to the main spent acid (particularly the formation of precipitates) to a distance of one to one and a half meters (3 to 5 ft) from the well. It must also assist clean up. Five percent hydrochloric acid or diesel oil is commonly used (with a volume along the same order of magnitude as the main treatment). For injection wells, the injection water can also be used directly.

It is of utmost importance to clean up the well immediately after the over/afterflush fluid has been injected in order to limit the risk of side reactions (with formation of precipitates, etc.) in the immediate vicinity of the wellbore. This is valid except for injection wells if water injection is begun immediately.

C. Additives

A large number of additives are available for use in making the acid itself more effective and, more important, in decreasing any adverse side effects. They must be determined according to the type of reservoir and the operating conditions by laboratory testing. Only really necessary additives should be used and the prescribed concentrations should be complied with. Check to see that they are all compatible with each other, with the treatment and formation fluids and with the formation itself.

The following additives can be mentioned:

- Corrosion inhibitors to protect equipment and limit the amount of metallic salts entrained in the acid. Their effectiveness depends on the temperature, the acid (type and concentration) and on the time. For very high temperatures, inhibitor fortifiers with a short-lived action are used in conjunction with corrosion inhibitors.
- Iron-retention additives, also called sequestering or complexing agents, are used to prevent the metallic salts in solution in the acid from precipitating out in the formation in the form of hydroxides when the acid is spent. It should be pointed out that, despite the use of a corrosion inhibitor, the acid will contain metallic salts, since inhibition is never 100% effective and the formation usually contains metallic minerals.
- Demulsifiers and antisludge agents to keep viscous products from being formed that are difficult to get rid of later on.
- Wetting agents to help the acid penetrate, then facilitate clearing out the spent acid, and improve the contact between the acid and the formation.
- Clay stabilizers to keep shales from swelling or dispersing and avoid the sizeable resulting permeability reduction.
- Surfactants that affect surface tension forces between fluids so that in-place fluids are readily displaced, the contact between the acid and the formation is improved and the final clean up is easier.

- Diverting agents: the injected acid has a natural tendency to penetrate the most permeable zones, whereas the desired effect is to act on the plugged zones instead. As a result, physical/chemical agents are used to divert the acid from the most permeable zones. They must also be easy to eliminate rapidly and completely. For example, polymer-type temporary blocking agents or, in perforated cased holes, elastomer coated balls (ball sealers), can be used. The ball sealers stick onto those perforations which take enough fluid and obstruct them temporarily.

D. Other treatment fluids used to eliminate plugging

Depending on the cause of plugging, acids may not always be suited, other fluids can be more effective.

For example:
- If there is an emulsion, demulsifiers in conjunction with a slug of diesel oil may give good results.
- If there is a salt deposit, the basic approach consists in flushing with water that may be heated.
- If there are paraffin deposits, in general there are two solutions: either chlorine-free solvents with the advantage that they are compatible with production, or more effective chlorinated solvents with the drawback that they can not be dispatched with the oil and/or gas since they are a poison for the catalysts used in refining.

When deposits are laid down in the tubing instead of in the formation, the wireline scraper technique can be used either alone or usually with solvents.

E. Pumping the treatment fluid down to formation

Whatever procedure is used it is important to comply with a number of points:
- All equipment must be tested at a pressure at least equal to the maximum pressure expected during the treatment.
- All equipment and installations (tank, pump, tubing, etc.) must be cleaned beforehand (washed with 5% hydrochloric acid and a corrosion inhibitor for example).
- Treatment fluids must be made up from clean fluids before injection, prescribed concentrations must be observed scrupulously for the different components (acids, additives, etc.), product quality must be checked (including the quality of mixing water) and fluid samples must be taken.
- The treatment fluid is injected into the formation at a "matrix rate" without exceeding the formation breakdown pressure.
- The well is cleaned up and the spent acid is removed as quickly as possible, as soon as the treatment fluid has been completely injected, particularly when sandstone formations are acidized, and here again samples must be taken.
- After the operation a precise treatment report is drafted, the pumping diagrams in particular are immediately completed with all relevant data.

The treatment can be performed before final equipment installation with a temporary string, for example a squeeze packer can be used (with a circulating sub) possibly along with a retrievable bridge plug. This allows the treatment fluid to be circulated down the well and as a result only a minimum volume is to be squeezed before the treatment fluid gets to the formation. However, it may be difficult afterwards to stabilize the well in order to pull out the temporary equipment. As a result, temporary plugging agents may be required and there is no guarantee they can be 100% eliminated later on.

When a washing tool or a squeeze packer together with a retrievable bridge plug is used, treatment can be carried out on an interval by interval basis. Otherwise, diverting agents may have to be used, especially if the treatment is to cover a considerable formation height.

It may be preferable to treat the formation after final equipment installation. Since the well is usually equipped with a tubing and a packer, it is therefore necessary to be able to squeeze all the tubing volume before the treatment fluid comes into contact with the formation. This is true unless a coiled tubing or snubbing unit is used (see Section 5.3). In many cases squeezing this volume proves impossible. Before starting to pump the treatment fluid, an injectivity test must of course be run to be sure injectivity is sufficient.

Whenever it is economically feasible, the best technique is to treat the formation after completion with a coiled tubing unit or, failing this, with a snubbing unit. These techniques provide the following advantages:

- the tubular can be cleaned by circulating before treatment
- the treatment fluid can be emplaced directly in front of all the zone that needs to be treated
- clean up can be boosted by injecting nitrogen if necessary
- the equipment used in the operation is pulled out under pressure without having to "kill" the well.

2.4.3.3 Hydraulic fracturing

A. *Principle and application*

Hydraulic fracturing is an operation to help improve fluid flow toward the wellbore. It consists in creating a permeable drain extending as far as possible into the formation after fracturing the rock. This process is used when the well flow rate is insufficient because of low natural matrix permeability (a few tens of millidarcys for oil reservoirs, even less for gas reservoirs) and not because plugging has occurred. The aim is to get enough conductivity contrast between the fracture and the formation.

Hydraulic fracturing is suitable only for properly consolidated formations (sandstone, carbonates) as opposed to plastic formations (shales, poorly consolidated sands). In addition, it is a highly unadvisable process if it might allow the unwanted inflow of a nearby fluid (if there is an interface).

In favorable cases productivity or injectivity gains stabilizing at around 3 to 4 can be achieved (excluding the unplugging effect).

B. Description of the process

There are five phases as described below.

a. Initiating the fracture

The pressure is raised in the well by pumping a fluid into it at a flow rate higher than what can leak off into the formation. Tensional stress is thereby generated and will initiate a fracture in the rock perpendicular to the minimum existing compressional stress before pumping.

b. Development or extension of the fracture

When pumping is kept up the fracture extends more and more as long as the pumping rate is greater than the leak off rate through the fracture faces.

At shallow depths, i.e. less than 600 m (2000 ft), the fracture usually develops on a horizontal plane (Fig. 2.26a) according to a more or less radial circular type of configuration. The fracture gradient (ratio between the fracture pressure on the formation and the depth) is about 23 kPa/m (1 psi/ft).

At greater depths, in particular deeper than 1000 m (3300 ft), the fracture normally develops on a vertical plane (Fig. 2.26b). In models, fractures are presumed to be either symmetrical in relation to the well or to develop on only one side of the well. The fracture gradient here is less than 23 kPa/m, and its average value is held to be 16 kPa/m (0.7 psi/ft).

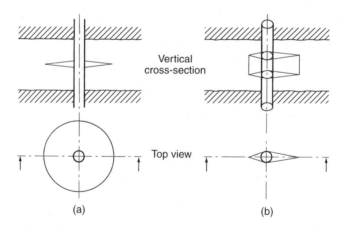

Fig. 2.26 Fracture configuration (*Source:* After IFP, Report No. 29255). a. Horizontal fracturing. b. Vertical fracturing.

c. Keeping the fracture open

There are two cases:
- In carbonate formations a fracturing fluid containing relatively concentrated hydrochloric acid is used. The acid etches the walls of the fracture irregularly, leaving residual

channels (worm holes) with very high permeability when the fracture closes back up once pumping has ceased. This is what is called acid frac.
- In sandstone formations, propping agents with very good permeability (sand, glass beads, etc.) are mixed with the fracturing fluid (usually a high viscosity fluid). They are carried along into the fracture by the fluid and keep the fracture from closing back up when pumping has been stopped at the end of the treatment. Propping agents should not be pumped in until the fracture has reached geometrical dimensions (fracture thickness especially) that allow them to pass.

d. Shutting in the well

This allows the excess pressure to be resorbed when the fracturing fluid leaks off through the walls of the fracture. It is crucial to shut in the well before the well is cleaned up when fracturing with propping agents so that the agents can be blocked in place before the well is cleaned up.

e. Cleaning up and bringing the well stream

Not only the treatment fluid in the residual fracture, but also the fluid that has leaked off into the formation must be eliminated.

C. Typical pressure diagram

The variations in pressure when a vertical fracture is being formed are shown schematically in Figure 2.27. During pumping, a pressure peak followed by a plateau phase can be distinguished. When pumping is stopped, it is followed by a dramatic pressure drop then by an approximately stable pressure.

The term fracturing pressure or breakdown pressure is used to describe the value seen on the pressure peak (it is the reference pressure). The plateau that follows corresponds to the propagation pressure. Then the final pressure at shut in is often called instantaneous shut in pressure, ISIP. It is often difficult to define with sufficient precision.

The pressures measured during the fracturing operation can vary considerably from this schematic presentation:
- The pressure peak is of variable amplitude and can even disappear.
- The propagation pressure always shows a slight variation, slightly upward or downward or even a series of the two trends.

D. Fracturing fluids

The main qualities required of a fracturing fluid are as follows:
- High viscosity and low filtration rate to get good fracture thickness and extension, and make sure the propping agents are properly emplaced (carrying capacity of heavy propping agents in high concentration).

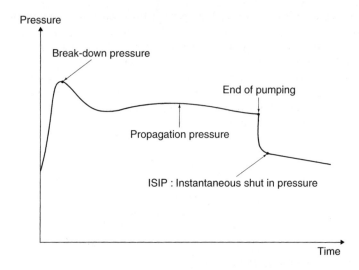

Fig. 2.27 Pressure variations during fracturing (vertical fracturing) (*Source:* ENSPM Formation Industrie).

- Low friction to reduce the pumping power required during injection.
- Good compatibility with the rock and the formation fluids, low insoluble solids content, and minimum production of insoluble reaction products so that the formation is not damaged.
- Readily displaceable by the reservoir oil and/or gas, low viscosity (during clean up) and low specific gravity to facilitate clean up and production start up.
- Suited to the temperatures prevailing during the operation (viscosity depends on temperature in addition to shear and duration).

These various properties are obtained by adding additives to the base fluid, the most conventional of which is water. Its viscosity is increased by producing a linear or cross-linked gel. The advantage of water is that it is a low-cost fluid that is readily available and transportable. It poses no safety problems (fire, explosion, pollution, etc.), is easy to treat with additives and its relatively high specific gravity (compared to degassed crude) means that the required pumping power can be reduced. However, for the same reason well clean up may be difficult if the reservoir pressure is low. Also, the relative permeability to oil has been lowered due to increased water saturation by filtration. It is important to pay attention to the quality of the water used (concentration in chlorides, sodium bicarbonate, iron, insoluble solids, bacteria, etc.).

To a lesser degree oil-base fracturing fluids (crude or gelled oils, emulsions) are used. The advantage of this type of fluid is that it is more compatible with the formation fluids and there is no solid residue. It has good stability, a good gel carrying capacity and its low specific gravity helps facilitate clean up. However, the cost is high, there are safety risks and more pumping power is required.

There are other fracturing fluids such as foams, liquid carbon dioxide gas and acids of course for carbonate formations (concentrated hydrochloric acid preceded by a water-base gel to initiate and develop the fracture, hydrochloric acid in emulsion in oil or gelled, etc.).

E. Additives

The same as for acidizing, a number of additives are often needed for fracturing. The type and concentration must be chosen carefully depending on reservoir parameters and on operating conditions. Here again laboratory testing is critical.

Let us mention the following in particular:
- Gelling agents used to increase the base fluid's viscosity and reduce its filtration rate and friction coefficient. They are usually polymers soluble in the base fluid (when the gelling agent has a slow hydration rate or when the well is shallow, care must be taken to prepare the gel in advance so that it can prehydrate before it gets to the reservoir, otherwise it can be prepared continuously during the operation).
- Cross-linkers designed to increase the gel's viscosity by transforming the gelling polymer's linear structure into a three-dimensional cross-linked structure.
- Fluid-loss additives that supplement the filtrate reducing action of the gelling agents (note that since fluid-loss additives are in fact plugging agents and usually insoluble, excess use may damage the formation).
- Friction reducers that supplement the friction reducing action of the gelling agents.
- Diverting agents, especially when multiple fracturing is planned.
- Breakers to break down the fracturing gel after a given residence time in the formation. These are either enzymes or acids for low temperatures, or chemical products if the temperature is higher than 80°C (180°F). The breaker concentration must take into account fracturing conditions as well as the scheduled time between gel preparation and injection.
- Surfactants that are sometimes needed to help the fracturing fluid flow back out of the formation after treatment.
- Other additives such as bactericides, pH control agents, antifoaming agents, etc.
- The various additives mentioned previously for acidizing treatments can also be used for acid fracturing.

F. Propping agents

Sand is the most common propping agent, mainly because it costs less than other products.

Also used are:
- Glass beads that provide better permeability than sand but tend to shatter into fine shards that plug up the fracture when their fracture stress is attained (40 to 50 MPa or 5800 to 7250 psi).
- What are termed high-strength materials (bauxite and zirconium oxide) with better mechanical strength (100 MPa or 14 500 psi) but whose higher specific gravity (3.7) may cause problems of emplacement in the fracture.

- What are called intermediate materials, that are designed to associate good strength and acceptable specific gravity.

In actual fact the most important property for propping agents is conductivity under bottomhole conditions, i.e. under stress, with temperature and in the presence of reservoir fluids. Since conductivity is the product of permeability times the thickness of the fracture, transportation and distribution of the proppant in the fracture are therefore also important parameters. Note that there may be a contradiction between certain required properties: steel balls that provide excellent compressional strength are too heavy to be transported properly.

Conductivity is more particularly dependent on the factors mentioned below.

a. Proppant grain size

Since the proppant grain has been sieved (little dispersion in the grain size, no fines, etc.), the permeability (several hundred darcys) is considerably greater than that of formations said to have low to very low permeability.

In addition, proppant permeability must be considered in relation to formation permeability. A very low permeability formation does not necessarily need very high permeability propping agents. In practice agents with 0.84 to 0.42 mm (0.033 to 0.017 in) diameters (20 to 40 mesh) are used.

b. Actual stress on the propping agent in the fracture

Figure 2.28 illustrates the effect on permeability of the stress effectively exerted on sand as a propping agent in the fracture after stimulation and putting the well back on stream.

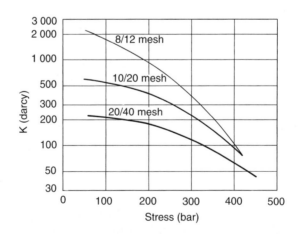

Fig. 2.28 Average permeability of sand versus the effective stress (*Source: Revue de l'Institut Français du Pétrole,* 42, No. 2, 1987).

c. Transporting proppants and filling the fracture

The carrying capacity depends on the viscosity of the carrier fluid, hence the use of gels, possibly cross-linked, and on proppant specific gravity.

Fracture filling also depends on the concentration in propping agent in the injected fluid, which is expressed by dividing the weight of the propping agents by the volume of liquid (and not of the slurry). In hydraulic fracturing a concentration range of 120 to 600 kg/m^3 (1 to 5 lb/gal) is fairly common, with 240 kg/m^3 (2 lb/gal) as an average value (for sand this represents a volume of propped up fracture of about 15% of the injected fluid volume). Higher concentrations (up to 18 lb/gal, i.e. a solid volume of 45% of the total volume) can sometimes be used.

d. Temperature, time and the type of reservoir fluids

Permeability decreases with temperature and time. This decrease, particularly versus time, depends on the type of in-place fluids. Although fresh water is detrimental, there is no rule for salt water, and oil and gas do not have any adverse effect.

G. Preparation and implementation

Hydraulic fracturing requires thorough planning at all levels of preparation and implementation.

a. Sizing hydraulic fracturing

Sizing the operation means aiming for the best possible productivity based on:
- the characteristics of the pay zone: productive thickness, total thickness, position of the cap rock, position of the water interface, permeability, oil, water and gas saturations, and results of production tests
- the status of the well and its equipment: tubing diameter, maximum allowable wellhead pressure, cement job quality, and the scale of any workover operations that might need to be done
- the materials and equipment needed or available before, during and after the operation: fluids, propping agents, tanks, blenders, pumps, well equipment and logging tools
- the cost of materials, equipment and services.

This is why numerical simulation models or other methods (reference to charts, prior experience, etc.) are used. The true result can not be found. This leads to simplification, but still gives orders of magnitude and shows their sensitivity to variations in operating parameters.

The main things to be determined are:
- the degree of confinement
- the propping agent injection plan (size, injected volume and concentration)

- the final dimensions (height, thickness and length of fracture propped up)
- the efficiency of the operation (ratio of the fracture volume over total injected volume)

and by means of a reservoir model:
- the gain in productivity
- the profitability of the operation (pay-out time).

The design of the operation is detailed in a fracturing program and schedule (see following points).

b. Defining the fracturing procedure

The operation is sized very early on, several months beforehand sometimes for large-scale operations since all required equipment and materials need to be brought in in good time.

When propping agent and fluid volumes and flow rates have been fairly well defined, it is then necessary to proceed to a detailed definition of the injection sequence.

There are three main phases:
- Injection of fluid without any propping agent (pad volume), which simply aims to develop the fracture for an acid frac, or in the other cases to open the fracture wide enough so that the propping agents can enter it (fracture thickness two or three times greater than the size of the largest grain injected). A pre-pad may be injected first to cool down the formation.
- Injection of acid or of slurry containing a propping agent; the slurry injection may include several phases with increasing solids concentration.
- Flushing which consists in injecting a volume of solids-free fluid that is equal to the tubing volume.

During the flushing phase care must be taken not to overdisplace the fluid containing the propping agent. Otherwise the near-wellbore, which is very important for fracture conductivity, will no longer be propped up. To prevent this, the sand-out or screen-out technique can be implemented. This consists in increasing propping agent concentration until refusal in the last injection sequences. Some sand then remains in the bottom of the hole and must be cleaned out.

When the formation is very thick or made up of permeable layers separated by large thicknesses of impermeable formation, it may become necessary to fracture several different levels in sequence after having isolated them from one another. They can be isolated mechanically (by using a packer and a bridge plug), or by using a sand plug or the limited entry technique (simultaneous fracturing of different levels using a high flow rate and a limited number of perforations in each zone).

c. Preparing the well

It is imperative for the production string, the casing and the bottom of the hole to be clean during the fracturing operation as such. The cleaning to be planned depends on the well's past history.

Another imperative is to be sure that the cement job is good on the levels adjacent to the target level. The connection between the wellbore and the formation must also be good, i.e. most of the perforations that are supposed to be open actually are open. When perforation plugging seems to be excessive, treatment is necessary prior to fracturing.

The tubing and the wellhead must be capable (or made capable) of withstanding the stresses caused by the fracturing operation. There are two types of stress: mechanical stress due to the fracture pressure and to pressure losses in the tubing, and thermal stress in deep or very hot wells.

The packer must be checked to be sure that it is capable of withstanding the fracture overpressure. In addition, in very deep or hot wells, the cooling effect of fracturing fluid flowing at high rates may cause tension or excessive movement in the tubing. These effects may be attenuated by moderately pressurizing the annulus.

d. Logging before fracturing

This involves what are termed production logs, i.e. performed with tools whose diameter is small enough so that they can be run in through the tubing. Performed before fracturing, they give a detailed characterization of the pay zone, i.e. they define the permeability contrasts between levels in the pay zone. However, the main advantage is to compare them with the same type of logs that will be run after fracturing.

e. Pre-job meeting

This meeting includes representatives of the operating company present during the operation (drilling engineer/supervisor, completion engineer/supervisor, production operator, safety officer), of the contractor and of service companies. It is held before the injectivity test and treatment in order to:
- Present the program.
- Define each player's role precisely and by name.
- Define maximum allowable pressures, safety and evacuation rules and regulations, emergency procedures in the event of trouble (leaks, failures, screen-out, etc.).

f. Injectivity test or pre-fracturing

The aim is to fracture the rock and then inject a volume of fluid which remains small compared to the volume that will be injected for the stimulation operation itself. Pre-fracturing is used to make sure that fracturing is possible. It gives an assessment of the fracture pressure required for sizing the operation. It also gives an approximate idea of the existing minimum horizontal stress and of the fracturing fluid filtration coefficient (a factor that is extremely difficult to determine in a laboratory). These values are indispensable in calculating the propagation of the fracture. The data obtained in this way may cast doubts on the fracturing operation as such.

g. Fracturing and measurements during fracturing

The equipment and materials brought in to the site include storage, blenders, pumping units, low and high pressure injection lines and flowlines.

The way the gel is made up (beforehand or during the operation) has an important effect on implementation. Fluid samples are taken before injection. The breaker, which is designed to break the viscosity, must obviously be used at the last minute. It is usually mixed with the fluid in the low pressure circuit, upstream from the injection pumps. If the gel is due to be cross-linked, the cross-linker is injected into the plain gel by means of a metering pump at the wellhead. When the fracturing fluid is due to include a specified amount of gas, the gas is injected continuously into the main line after the gel and propping agent have been mixed. The propping agent is mixed with the gel upstream from the injection pumps. Dispersing it homogeneously throughout the gel is still a problem that needs more scrutiny.

Equipment type, number and cleanness, and product quality and quantity have to be checked prior to the operation. Likewise the installations have to be pressure tested beforehand.

In hydraulic fracturing operations, measurements are crucial in order to:

- Monitor the operation and be sure that the treatment proceeds as planned.
- Act on the process to change a parameter or even stop the operation.
- Provide the means to analyze the operation afterwards and draw conclusions about the quality of its design and the modifications that should be made for operations on other wells.

The measurements made during fracturing are today becoming increasingly refined; they are benefitting from progress achieved in digitizing and processing the data supplied by sensors.

The major measurements are wellhead pressure, annular pressure, flow rate, volume and specific gravity. Specific gravity is an important measurement since it is the only way to determine and monitor the propping agent's dispersion in the fracturing fluid. According to the specific gravity readings, the propping agent hopper feed valve is operated so as to get as close as possible to the concentration that is believed to be ideal.

At the end of the operation the residual contents of tanks, bins, etc. must be assessed to check and recalibrate the measurements made during the operation (flow rate, etc.).

h. Measurements after fracturing

These are measurements made immediately after the operation or a short while later. The fact is that measuring and interpreting flow rates in the long term is more specifically a reservoir management activity.

It is useful to record the pressure once injection has been stopped (see Section j., Monitoring the results of the operation). The volume of fracturing fluid cleared from the well is a useful indication in assessing the fluid's behavior (leak off, breaking after the operation, makeup problems). Production logs can be run in order to assess the fractured height.

i. Putting the well back on stream

The well is put back on stream taking two contradictory conditions into account:

- The fracture faces must exert sufficient stress on the propping agent so that it is not carried off into the well when the fracturing fluid is cleared. As a result, there must be a waiting phase once injection has ceased in order to let the fracturing fluid leak off through the fracture faces.
- The waiting phase must, however, be limited to the minimum time required so that the formation's permeability to the reservoir fluid is not affected. Problems of sand sedimentation in the well must also be obviated in the event of screen-out (the best solution is to eliminate the sand under pressure with a coiled tubing or snubbing workover unit).

A gas may be incorporated into the fracturing fluid in order to facilitate clearing. The gas is usually nitrogen and its expansion in the production casing causes the bottomhole backpressure to decrease (gas lift effect).

Clearing is performed under choke control and at reduced flow rate, while monitoring that no sand comes back out (except in the event of screen-out). Samples of the cleared fluid are taken regularly and production parameters such as those listed below are measured:

- wellhead pressure
- flow rate
- effluent specific gravity
- BSW (basic sediment and water: the percentage of sediment and water in the effluent)
- GOR (gas oil ratio).

j. Monitoring the results of the operation

The flow rate measurement is the most important, with a distinction made between the oil, gas and water flows. There can be a wide variety of results which run the gamut between complete success and absolute failure, depending on the increase in target fluid (oil or gas) flow rate and any increase in unwanted fluid (water, gas in some cases) flow rate.

A satisfactory operation means that the flow rate is at least doubled in the long run, with the world average situated at around 3.5. This of course excludes any transient or unplugging effects. The post-fracturing flow rate does not give enough information to understand what happened during the operation. Ideally, the following points need to be known:

- during fracturing: the fracture's dimensions, i.e. height, thickness and length
- after fracturing: the same factors as well as the permeability of the emplaced propping agent.

The present status of this type of measurements is now as follows:

- the only factor that can be determined on an industrial scale is the height of the fracture, assessed on the basis of certain logs (temperature logging, flow metering, and nuclear log when the propping agent has a radioactive tracer)
- methods that are still more or less experimental are used to determine the direction of the fracture and even its length and height far from the well (electric, acoustic, etc. methods)

- an attempt is made to deduce information about the height of the fracture, its length, its conductivity and about fracturing fluid filtration (methods developed after Nolte's research in particular) from the fracture pressure and the variation in pressure after injection.

H. Some orders of magnitude

Table 2.2 summarizes some orders of magnitude for vertical fractures.

Table 2.2 Orders of magnitude for vertical fractures (*Source:* ENSPM Formation Industrie).

	Unit	Conventional vertical fractures	Massive frac
Pad volume	m^3	30 to 100	≈ 1000
	gal	8000 to 25 000	≈ 250 000
Slurry	m^3	50 to 200	≈ 1000
	gal	12 000 to 50 000	≈ 250 000
Average concentration	kg/m^3	300 to 420	idem
	lb/gal	2.5 to 3.5	idem
Propping agent	10^3 N	150 to 800	≈ 4000
	10^3 lb	30 to 160	≈ 800
Average flow rate	l/min	800 to 3200	2400 to 4800
	bpm	5 to 20	15 to 30
Pumping time	h	1 to 4	8 to 15
	h	1 to 10	8 to 15
Pumping power[1]	kW	some 1000	> 7500
	HP	some 1000	> 10 000
Fracture thickness[2]	mm	5 to 20	
	in	2 to 8	
Extension from the well axis[2]	m	50 to 400	≈ 1000
	ft	150 to 1200	≈ 3000
Height[2]	m	10 to 1000	
	ft	30 to 300	

1. The required pumping power depends on the depth at which the fracture is created and on the pumping rate.
2. The corresponding values are highly variable depending on the characteristics of the formation and of the pumped fluid.

Nota bene: In the case of horizontal fractures, which are much less common, the fracture thickness is only a few millimeters and the extension (radius) is a few tens of meters.

2.5 THE SPECIAL CASE OF HORIZONTAL WELLS

Horizontal wells are not a new concept, since as far back as in the fifties and sixties experimenting was already being done on drilling this type of well in the USSR especially (the trials were fairly disappointing though in terms of productivity). In fact it was in the late seventies and early eighties that the technique became really popular, carried along on the wave of progress in directional drilling and the need for new oil supplies.

Horizontal wells may provide a number of advantages in exploiting reservoirs. However, they also pose a whole series of specific problems. To solve them, traditional techniques had to be adapted and new techniques invented. It is a field still in full development.

Only those problems specific to the connection between the borehole and the pay zone will be dealt with here. For those involved in horizontal drilling, refer to documentation on the subject, and for those involved in operations on horizontal wells, see Chapter 5, Section 5.5.1.

Before looking at these advantages and specific problems, it would perhaps be advisable to define exactly what is meant by horizontal wells. They are wells that have an inclination inside the reservoir greater than 80° or 85° with respect to vertical. Today such wells quite commonly have horizontal drain holes that are 2000 to 3000 meters long (6000 to 10 000 ft).

2.5.1 Advantages in producing reservoirs

Depending on reservoir configuration and provided the configuration is well known before undertaking horizontal drilling, horizontal wells can:

- increase the recovery rate, mainly due to a greatly improved productivity index, and/or
- boost the recovery ratio, and/or
- help solve certain production problems.

The main cases are presented below.

2.5.1.1 Low permeability formation

In this type of reservoir, hydraulic fracturing may be performed but (in comparison with a horizontal well):

- Fracturing is much less extensive.
- The residual permeability of the fracture is much lower.
- The orientation of the fracture is not controlled.
- The vertical extension of the fracture is hard to control, the fracture may extend out of the formation or afford access to a zone where there is an unwanted fluid: aquifer, gas cap.

A horizontal well does not have all these drawbacks. However, the fluid must flow toward the horizontal drain hole not only on a horizontal plane (the same as for vertical wells or hydraulically fractured wells), but also on a vertical plane. This may be detrimental:
- if the reservoir is thick (this is more or less compensated with multi-drains)
- if the vertical permeability is much lower than the horizontal permeability.

Except for these cases, a horizontal well gives a much better productivity index for a low-permeability formation than a vertical well. To a lesser degree it is also better than a hydraulically fractured well.

However, this effect becomes partly attenuated with time, as the produced fluid comes from reservoir zones that are farther and farther away from the horizontal drain hole.

The improvement in the productivity index (*PI* multiplied by 2 to 4, even 10 if $k_v = k_h$) means:
- faster recovery, or
- fewer wells, or
- a lower pressure differential $(P_R - P_{BH})$(which, as we will see later, may help solve certain production problems).

2.5.1.2 Thin formation

Here a vertical well's productivity index is limited by the small contact area between the well and the producing layer. A horizontal well can offset this (provided the vertical permeability is not too low compared to the horizontal permeability).

2.5.1.3 Plugged formation

Considering again that the vertical permeability is not too low compared to the horizontal one, the plugging effect is attenuated by the length of the drain hole. Nevertheless, plugging will lower the flow capacity, compared to the theoretical one. So horizontal drain cannot be considered as a cure to plugging.

2.5.1.4 Effect of turbulence

A high flow rate in the near-wellbore area, a condition that is often encountered in gas wells but also sometimes in oil wells, causes turbulence which in turn causes further pressure losses (as much as 30% and more of the total pressure losses).

Because of the length of the horizontal drain hole, the flow rate is reduced inside the formation. This lessens the turbulence effect, sometimes even making it negligible. As a result, the well's potential is improved, especially for gas reservoirs.

2.5.1.5 Critical flow rate (in relationship with coning)

In some reservoirs the wells must be produced at a rate lower than a critical flow rate. Above this lower rate an unwanted fluid (water, gas) is sucked into the well by a coning effect.

This is the case for the following:

- A gas reservoir with an underlying aquifer (a slight water inflow can be highly detrimental in terms of back pressure on the producing layer).
- An oil reservoir with a gas cap (producing gas means losing part of the energy that enables the oil to be produced).
- An oil reservoir with an underlying aquifer (unnecessary depletion due to production of water, increase in back pressure on the producing layer, unwanted water on the surface).

Horizontal wells offer a double-barreled advantage:

- A better productivity index which allows the pressure differential on the producing layer to be reduced, thereby limiting the influx of unwanted fluid.
- The drain hole can be kept at a maximum distance from the interface between the target fluid and the unwanted fluid.

The outcome is a critical flow rate that may be several times greater (from 2 to 4 times) than that of a vertical well, and therefore gives faster recovery. An additional asset is also a better recovery ratio since the unwanted fluid remains in the formation.

2.5.1.6 Insufficiently consolidated formation

In these formations sanding up is due to viscosity forces and therefore to the fluid velocity among other causes. Because of its length the horizontal drain hole reduces velocity, and as a result the tendency to carry sand grains along with the flow. However, there is no means today of determining beforehand the critical velocity that should not be exceeded.

Moreover, a lot more sand can accumulate in the drain hole before it gets plugged up.

Finally, when sand control by sand screen is implemented, the sand control lifetime can be expected to be longer. Plugging occurs less quickly (lower velocity reducing fine particle entrainment, larger inlet area) and has less effect on the productivity (longer drain hole).

2.5.1.7 Naturally fractured, heterogeneous formation, etc.

Natural fractures generally develop in a vertical plane. The horizontal well is therefore the best way to intercept the largest number of these fractures and take advantage of their drainage capacity (provided the orientation of the fractures is known). Likewise in cases of heterogeneous, multilayer, etc. reservoirs the horizontal well has a better chance of finding the largest number of zones with good characteristics.

Once again, all the cases mentioned mean an improved productivity index, and the increase can be considerable (the PI can be multiplied by 10 or even much more).

2.5.1.8 Secondary recovery

Horizontal wells can enhance the efficiency of secondary recovery:
- better injection capacity
- injection is better distributed throughout the reservoir causing less interface deformation, less fingering and therefore slower breakthrough and better sweeping efficiency.

2.5.2 Problems specific to the pay zone-borehole connection
2.5.2.1 Pay zone-borehole connection configuration

Numerous existing horizontal wells are equipped with a pre-perforated liner but, all the configurations are possible (open hole or open hole plus pre-perforated liner, pre-perforated liner partially cemented or with inflatable packers, liner cemented and then perforated).

One may think that the ideal horizontal drain hole that intercepts only one reservoir and one fluid does not theoretically require anything more than a pre-perforated liner and may even be left in open hole completion (if the geomechanical properties of the formation are sufficient), combined with a sand control process, if necessary, but this may not be suitable if an acidizing or fraturing job is required.

All the more so, wells that have penetrated major heterogeneities, different facies, faults or have crossed several reservoirs, a gas cap, or a layer connected to a water drive may require more suitable completion in order to isolate certain zones from each other. There are two means of achieving this: cementing and inflatable packers.

Inflatable casing packers are commonly used in existing horizontal wells. The main purposes are: isolating intermediate zones, for example to close a water-producing fault or a gas-bearing zone; or dividing up the drain hole into several stretches to deal with each zone selectively independently from the others.

Cementing should be confined to specific cases. Even though the operation poses technical problems, it is feasible (along with the relevant perforating). However, the length to be cemented is considerable. This means long time frames and particularly high operating costs, which may become excessive.

In actual fact the liner will be cemented when:
- Hydraulic fracturing is due to be performed.
- The non horizontal part of the well needs to be protected (going through a gas cap for example), in this case partial cementing is enough.

Additionally, choosing the type of pay zone-borehole connection is compounded by the fact that it depends not only on initial data but also on the way certain bottomhole parameters vary with time. However, it is difficult, for example, to anticipate where the breakthrough of an unwanted fluid (water, gas) will occur in a horizontal well. As a result it is important to choose an initial configuration that may be adapted later on according to the well's behavior.

2.5.2.2 Running in and cementing the liner

No particular difficulty is involved in running the liner in, except for centering. The number of centralizer should not be so large that it impedes displacement, but it must be large enough to help slide the liner in (it acts like a skid). Also, when the liner is cemented it must allow a proper thickness of the cement sheath.

In the same way as for a vertical well, the important point about cementing is for the annular space to be properly filled up and for the cement to bond well with the formation and the pipe. This will give a good seal and therefore good selectivity.

To achieve this, it is necessary to:

- Clean the horizontal part of the borehole thoroughly and at the same time get the cuttings out of the vertical part (this means alternating different circulation sequences).
- Center the pipe to keep at least 2.5 cm (1 in) of clearance.
- Use a large volume of suitable lead spacer fluid.
- Improve the cement slurry composition to prevent any migration of water which can cause free-water channeling along the top of the drain hole when the cement sets.
- If possible pump the cement in under turbulent flow conditions (if not, increase the volume used and displace the slurry under plug flow, or slow flow, conditions).

2.5.2.3 Perforating

First of all, remember that since the horizontal extension is long, the share of perforating costs in the final cost of the well may be very high (the shallower the well, the higher the share). This may even lead to second thoughts about the advantages of having a cemented liner. The traditional wireline perforating methods can not be used, but all the methods for manipulating logging tools in a horizontal borehole can be applied (see Chapter 5, Section 5.5.1).

TCP can of course be used, but in this case:

- The guns can not possibly be left in the bottom of the hole, and will have to be pulled back out before running the downhole equipment.
- The guns are restricted by their capacity to "turn the corner" if the well has a small or medium curve radius.

Especially in poorly consolidated formations, the important parameter is not to perforate underbalanced with the largest possible charges, but rather to keep from getting stuck. To achieve this, it is preferable to:

- Limit the underbalance or even perforate overbalanced.
- Use guns with a small enough diameter to allow for a washover operation.
- Use centralizers and a quick-release system.

Guns with charges limited to a specific angular domain (for example a 60 to 180° angle facing downward) can be used to keep from having too much clearance (when the gun diameter is small in relation to the casing diameter) or to help control sand by gravel packing. However, although there are methods to determine the orientation of the top of the gun, how can the whole gun be turned to reorient it, since rotation at the wellhead will generate a torsional movement in the gun due to friction?

2.5.2.4 Sand control

Less sand influx can be expected in a horizontal well than in a vertical well. The horizontal drain hole causes less coning and thereby delays water inflow. It also considerably slows down velocity near the wellbore due to its large surface area. However, since the sand entrainment threshold is very low, this does not solve the problem of sand control (even if the sand flow into the wellbore is low, it will accumulate and ultimately cause production problems).

Consolidation methods do not really seem applicable because:
- just one untreated zone can make the process fail
- treatment products need to be squeezed into the formation

and the horizontal drain hole has a very large surface area, with non-homogeneous characteristics in addition.

Although gravel packing is usually the most effective sand control process in vertical or deviated wells, the same does not hold true for horizontal wells. The effectiveness of gravel packing is in fact closely related to a large mass of gravel all around the sand screens. This requires the screens to be properly centered and also means that the space between them and the walls of the hole must be properly filled with well-compacted gravel. Gravity plays a crucial role in these parameters. In spite of present-day procedures, it may be difficult to obtain this in horizontal boreholes.

A perforated (or pre-perforated) liner with openings only along the bottom would facilitate things, but it is difficult to achieve proper orientation over such a long distance. Additionally, when the gravel is set into place there are problems of carrier fluid filtration — the longer the drain hole, the greater the problems. Filtration causes increased concentration in gravel and slower carrying velocity, all of which can cause screen out.

Therefore, the screens alone sand control process suits well to horizontal wells. The type of screen (wire wrapped, reinforced, pre-packed, etc.) should also be chosen with due consideration to running-in conditions (curve radius, compression when the screens are pushed along the drain hole, etc.).

In order to facilitate setting the screens in place and make future operations (workover, etc.) easier, it is recommended to run them in:
- equipped with centralizers
- through an abundantly pre-perforated liner rather than in an open hole.

Note that:
- The usual rules are used to determine screen slots and the gravel size (in the gravel pack case).
- Since the drain hole is so long, the price of equipment is high (a meter of sand screen costs several hundred dollars).

2.5.2.5 Stimulation

Acidizing and fracturing as such are applicable to horizontal wells.

In acidizing operations:
- It may be difficult to ensure treatment selectivity depending on the type of connection between the borehole and the pay zone, or time-consuming methods may have to be used.
- The use of diverting agents such as ball sealers is to be avoided, because they are hard to get rid of and their accumulation in the drain hole may jeopardize future operations (measurements made in the drain hole for example).

As for fracturing operations, whether with acid or with propping agents:
- The operation is mainly designed for wells with cemented liners (in wells with non-cemented liners, but with the different zones staked out with packers, it is not usually possible to increase the pressure high enough; only acid washing or low-pressure squeeze can be performed).
- It is useful to know in situ stresses, since the fracture's orientation and therefore position in relation to the horizontal drain hole depend on them.

2.5.2.6 Configuration of production string(s)

Horizontal drain holes are often produced though a single zone completion. There is also more and more selective completions (Figs. 2.29 and 2.30) where circulating sleeves allow to:
- isolate the selection(s) (in case of a single drain well) or the drains (in case of a multi-drains well) which start to produce an undesirable fluid (water, gas);
- adjust more or less the flowrate of the various sections (case of a single drain well) or the various drains (case of a multidrains well), depending on the position of the circulating sleeve (provided the circulating sleeve design allows it);
- treat preferentially such or such section or drain;
- open more or less such or such section or drain which produce more gas to make lighter the weight of fluid column in the tubing and so to increase the oil flowrate procured by the other sections or drain (self gas-lift).

Depending on the design, the circulation sleeve can be actuated:
- mechanically with a tool run at the end of a small-diameter tubular (a coiled tubing unit for example),
- hydraulically or electro-hydraulically though one or several hydraulic lines associated, if the case arises, to an electric line allowing to transmit a given number of information (temperature, pressure inside the sleeve, pressure outside the sleeve, sleeve position, etc.) and eventually to select the valve to be hydraulically actuated; the system can also be totally electrically actuated.

There is also parallel multi-tubing string completion and more specifically dual tubing string (Fig. 2.31) which can also be combined with alternate selective completions.

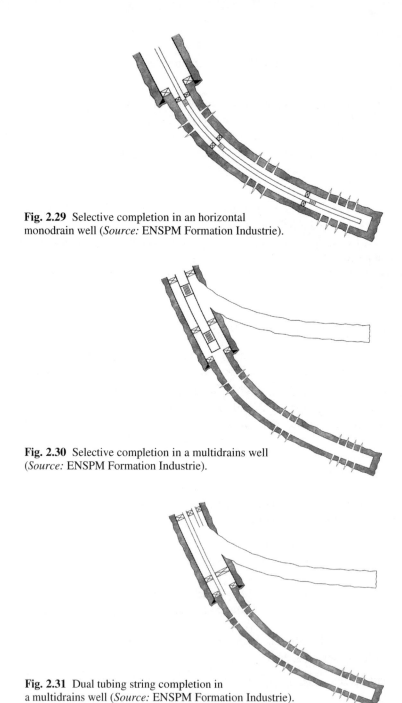

Fig. 2.29 Selective completion in an horizontal monodrain well (*Source:* ENSPM Formation Industrie).

Fig. 2.30 Selective completion in a multidrains well (*Source:* ENSPM Formation Industrie).

Fig. 2.31 Dual tubing string completion in a multidrains well (*Source:* ENSPM Formation Industrie).

Chapter 3

THE EQUIPMENT OF NATURALLY FLOWING WELLS

In well architecture there is a wide range of equipment available to get the effluent from the reservoir up to the surface where it will be treated. In searching for the best compromise between reservoir versus production requirements and constraints, the choices are governed by four main principles:

- access to the reservoir for measurements
- efficient transfer of the effluents from the bottom to the surface
- control of the effluents on the surface
- safety of all facilities.

The various equipment components are of course selected depending on the data collected, the configurations determined (connection between the borehole and the pay zone, single- or multiple-zone completion, etc.) and any artificial lift due to be performed. As a result, only the general criteria for choice will be given here. It should also be noted that the solution that is eventually chosen may be different from what was originally planned from a technical standpoint because the selected equipment:

- is not available on the market or can not be obtained in the specified time limit
- is excessively expensive
- is different from the equipment traditionally used by the company
- can be replaced (more or less satisfactorily) by leftover equipment in company stocks.

3.1 GENERAL CONFIGURATION OF FLOWING WELL EQUIPMENT

From top to bottom (Fig. 3.1) flowing wells usually include the following production equipment:

- A production wellhead with the Christmas tree and the tubing head.
  The Christmas tree comprises a series of valves, a choke and connections. It provides a means of controlling the effluents, ensuring the safety of the facilities and giving measurement tools and instruments access to the well.

3. THE EQUIPMENT OF NATURALLY FLOWING WELLS

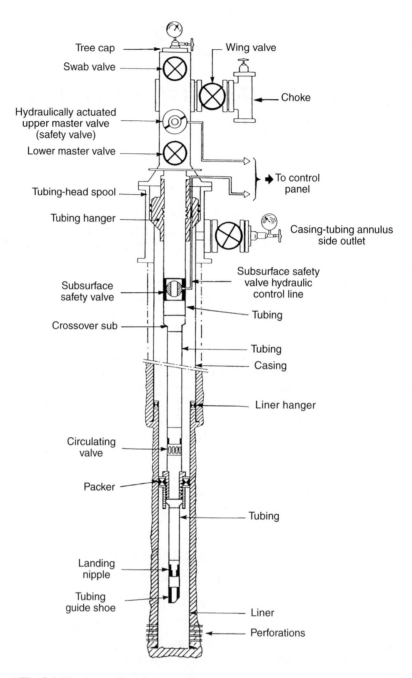

Fig. 3.1 Flowing well, one pay zone (*Source:* ENSPM Formation Industrie).

The tubing head or tubing-head spool, accommodates the device designed to hang the tubing(s).
- The tubing, pipe to carry the effluents from the bottom of the well up to the surface. Choosing the right pipe steel and through diameter contributes to the safety of the facilities and ensures that the effluents will reach the surface as efficiently as possible.
- An annular seal or production packer, which is used first and foremost to isolate the casing from the pressure in the well and from physical contact with the effluents which are sometimes highly corrosive.

  The choice of packer type has a considerable impact on running and setting procedures but also on the techniques and procedures used for further work on the well in the future.
- Downhole accessories such as sliding sleeve valves and landing nipples (tubing parts with a specially designed inside profile).

  These components are incorporated into the tubing. They allow circulation between the tubing and the annulus, or are used for installing equipment or to make it easier to use measurement and maintenance tools.

  A good rule is to limit the number of accessories to the ones that are strictly necessary for equipment setting, maintenance and workover operations.
- An extra safety valve, called a subsurface safety valve, for high-risk wells (offshore, subsea, gas-producers).

  It is designed to offset any failure of the Christmas tree safety valves or of the wellhead itself.

  This supplementary safety valve is incorporated in the tubing and is located approximately 30 to 50 meters (100 to 150 feet) below the ground level onshore or the bottom of the sea (mudline) offshore. It is controlled from a control panel on the surface via a hydraulic line.

For more specific requirements, other equipment may be used such as:
- perforated pipe
- flow couplings (reinforced couplings)
- blast joints (reinforced pipes)
- safety joints
- slip joints
- disconnection joints (dividers).

3.2 THE PRODUCTION WELLHEAD

The tubing must be hung and secured on the surface, and it needs a stack of valves and other accessories on top of it to meet safety and fluid flow requirements (Fig. 3.2). The choice of the type of wellhead and the functions it has to fulfill are related to the following requirements and needs:

- protection against uncontrolled flow from the well
- well flow rate control (choking)
- periodic monitoring of the well status and/or placing the well in safe condition by wireline tools run into the well
- withstanding pressure and temperature during production, when the well is shut in or during exceptional operations (hydraulic fracturing, for example).

3.2.1 Hanging (and securing) the tubing (Fig. 3.2)

The tubing head rests on the upper flange of the last casing head or is sometimes directly bored in the upper part of this casing head (compact heads). It accommodates the tubing hanger. The inside profile of the tubing head spool also provides a seal between the casing-tubing annulus and the top of the tubing hanger when the hanger is set in place.

The two side outlets that communicate with the annulus are closed by valves. One of these outlets is equipped with a pressure gage to observe the annulus and monitor any anomalous rise or fall in pressure. The other outlet is for connecting to outside piping.

The tubing hanger is usually machined to accommodate a back-pressure valve, BPV, during operations on the wellhead. Present-day tubing hangers usually have an extended neck with additional seals. Moreover, if the well is equipped with a subsurface safety valve a port is provided for control line hydraulic continuity.

3.2.2 The Christmas tree (Xmas tree) (Fig. 3.2)

It generally includes (from bottom to top):
- (one or) two master valves
- a cross
- a swab valve
- a tree cap.

All of this is completed by one or two wing valves and a choke. This set up allows:
- tools to be run in directly in line with the well provided a lubricator is screwed onto the tree cap connector
- the well to be opened or shut via the wing valve
- the flow rate to be controlled and adjusted via the choke bean
- the well to be placed in safe condition via the master valves.

The number of valves and their location is not fixed and must be adapted to the safety and production requirements specific to each field. High pressures require two or even three master valves exceptionally, high flow rates mandate installing wing valves on both lateral outlets, and lastly some cases demand continuous use of one or both outlets on the annulus.

3. THE EQUIPMENT OF NATURALLY FLOWING WELLS

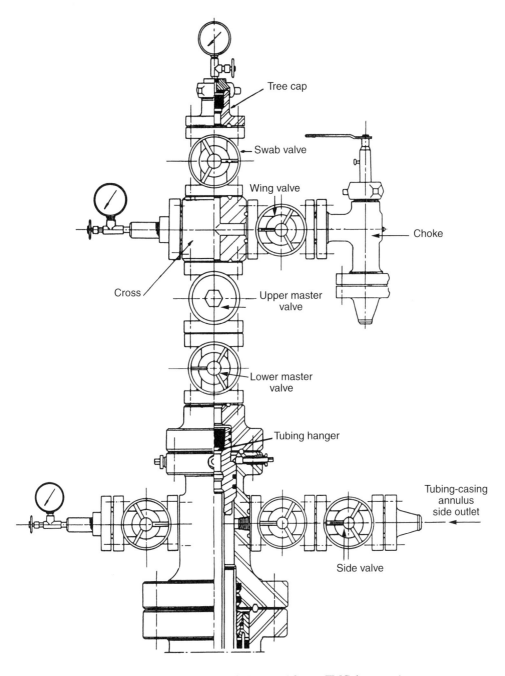

Fig. 3.2 Production wellhead (*Source*: After an FMC document).

In order to save space, a particularly critical point on offshore production facilities, solid block production wellheads are used more and more. Here master and swab valve seats and seal bores are directly machined in.

The valves used on the Christmas tree are of the gate valve type (Fig. 3.3).

During normal production, the lower master valve is kept open. The upper one is used to place the well in safe condition, it is closed automatically by means of a hydraulic or pneumatic control system (Fig. 3.4).

The control system for this surface safety valve, SSV, consists schematically speaking of a piston and a return spring. It is designed so that the valve is opened by pressurizing the control piston. At the end of the stroke, the valve is held open by maintaining the pressure on the control piston. Hydraulic slide valves keep the pressure on the control system as long as the parameters downstream from the valve are satisfactory.

However, as soon as the wellhead pressure comes out of the range authorized in normal production, the pilot valves switch over and cause pressure to bleed off in the control system. The return spring causes the valve to close. This is characteristic of a valve in the safety position (here closed) when there is no control signal (here the pressure on the piston of the control system), and is termed fail safe.

The wing valve, often manual, can be motorized so that it can be remote controlled.

Apart from safety system tripping, the well is closed by acting on the wing valve then on the upper master valve. In contrast, the well is brought back on stream by opening the upper master valve then the wing valve in order to "save" the upper master valve which is much more complicated and costly to replace than the wing valve.

When equipment is chosen, the inside through diameters, working pressure, metallurgy and flowhead configuration are given particular consideration:

- The inside vertical diameter must be at least the same as that of the tubing, the diameter of the wing valves being in proportion with those of the tubing and the flowlines.
- The rated Working Pressure (WP) is chosen depending on the highest pressure that could occur during the lifetime of the well, usually the shut-in wellhead pressure or the pressure prevailing during special operations. Rated working pressure found on the market are 5000 psi WP (35 MPa), 10000 psi WP (70 MPa) and now 15000 psi WP (100 MPa) as per standards established by the American Petroleum Institute (API).
- The metallurgy and type of seals depend on the type of effluent and its temperature, the fire resistance and working pressure.
- Choosing the configuration as such depends on safety conditions, floor space and production requirements.

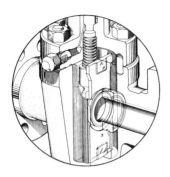

Fig. 3.3 Cameron type F valve (*Source:* Cameron).

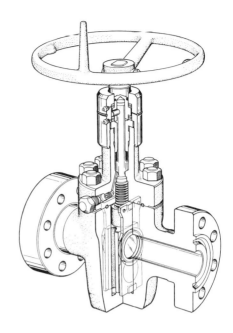

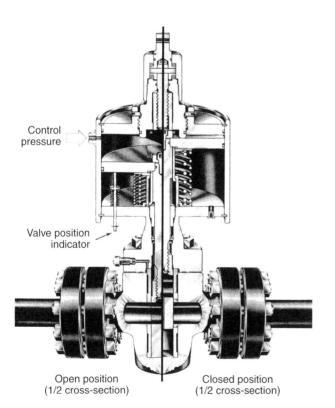

Control pressure

Valve position indicator

Open position (1/2 cross-section)

Closed position (1/2 cross-section)

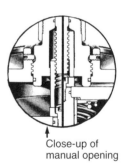

Close-up of manual opening

Fig. 3.4 Cameron automatic safety valve (*Source:* After a Cameron document).

3.3 THE PRODUCTION STRING OR TUBING

The production string (or tubing) is the pipe that carries the effluents from the reservoir to the surface in producer wells and from the surface to the reservoir in injection wells. Except in very specific cases, the tubing is utilized in conjunction with a packer, and this helps protect the casing from the effluent. The best fluid flow is achieved by choosing the right inside through diameter for the pipe, thereby minimizing energy consumption.

A tubing may have to be replaced in order to repair or adapt it to a change in production parameters, such as flow rate. In this sense, the tubing should also be considered as one of the consumables.

In conjunction with a circulating device, the tubing serves when necessary to displace fluids of required specific gravity and physicochemical composition in the well and/or annulus to meet start up or well control and casing protection needs.

If the diameter governs the flow rate, the steel metallurgy and type of between-pipe connection give the tubing a certain resistance against the effluents' chemical aggression and physical state (liquid or gas). Lastly the grades of steel used and the thickness enable the pipe to withstand sometimes high mechanical stress and also contribute to its corrosion resistance.

The pipe chosen to complete the well is the best trade-off between production constraints, well parameters, and sometimes efficient management of an existing stock of tubular goods.

3.3.1 Tubing characteristics

The tubing is made of drawn, seamless pipe which is usually fitted with a coupling. In addition to a traditional range listed in API schedules, manufacturers and vendors propose pipe made from improved steels or special steels. This special purpose pipe is used to meet the particular corrosion resistance requirements due to the presence of hydrogen sulfide or carbon dioxide and water. Parameters defined by API are mainly: nominal diameter, nominal weight, grade of steel, type of connection and length range.

3.3.1.1 Nominal diameter and geometrical characteristics

A. Nominal diameter

This is the outside diameter of the pipe body, or OD, traditionally expressed in inches and fractions of an inch. The following are standard outside diameters:

in	1.315	1.660	1.900	2 3/8	2 7/8	3 1/2	4	4 1/2
mm	33	42	48	60	73	89	102	114

Note that there is no reason why heavy producers can not use pipe manufactured according to casing standards (5", 5 1/2", and 7" in particular) to fulfill the tubing function, hence the expression: "The tubing of this well is a 5" casing". On the contrary, tubings with a diameter under 2 3/8" are mostly reserved for operations on the well using concentric pipe (snubbing operations) and are termed macaroni strings.

B. Inside diameter and thickness

The inside diameter, ID, is a result of the nominal diameter and the thickness of the pipe and is naturally the diameter used in calculating pressure losses and velocities.

C. Drift diameter

This parameter is of foremost importance. It is the guaranteed through diameter for a mandrel containing a specified cylindrical portion (a cylindrical portion with an outside diameter equal to the drift diameter and a length of 42 in or 1.06 m). In other words, the drift governs the range of equipment that can be run through the tubing whether for wireline tools, perforators, logging tools or small concentric tubings. As a result, the pipe run in the well, as well as the accessories, must absolutely be checked to be sure they conform to the drift.

D. Maximum outside diameter

This corresponds to the outside diameter, OD, of the connection and depends on the nominal diameter and on the type of connection. It is a critical parameter when there are problems of space in the casing. (For example, large diameter tubing, multiple string completions, etc.)

E. Pipe length

Because of the type of machining, each pipe naturally has a specific length. Pipe is classified into two length ranges:
- range 1: from 20 ft to 24 ft (6.10 m to 7.32 m)
- range 2: from 28 ft to 32 ft (8.53 m to 9.75 m).

The length range for the site must be chosen with care. It must be compatible with the hoisting and storage capacity of the mast of the rig used when the equipment is run in during completion as well as later on during workover operations. For production string spacing out operations there is also a range of short pipe, or pup joints. API lists various lengths for these joints: 2, 4, 6, 8, 10 and 12 ft (i.e. 0.61 m, 1.22 m, 1.83 m, 2.44 m, 3.05 m, and 3.66 m respectively).

3.3.1.2 Nominal weight

In addition to the nominal diameter, pipe is characterized in practice by its nominal weight, not its thickness. This corresponds to the average weight of a length of pipe, connections included, and is expressed in pounds per foot (lb/ft or the symbol #). Pipe with different nominal weights can be found with the same nominal diameter depending on the thickness. In contrast, slight variations in the nominal weight (Table 3.1) indicate different types of connections (see Section 3.3.1.4.A).

Table 3.1 Nominal weights for two thicknesses and two types of API connection for a 3 1/2" tubing.

Thickness		Nominal weight (#)	
(mm)	(in)	with API NU (or VAM) connection	with API EU connection
6.45	0.254	9.20	9.30
9.52	0.375	12.70	12.95

3.3.1.3 Grades of steel and metallurgical characteristics

Manufacturers supply standard API grades of steel, "proprietary" grades beyond standards and special steel for severe corrosion, particularly due to hydrogen sulfide or carbon dioxide. Note that the National Association of Corrosion Engineers (NACE) publishes standards that equipment must comply with to be considered "H_2S service".

A. API standard steels and grades for tubings

The following grades are included in API schedules: H40, J55, C75, L80, N80, C90, P105. However, they are not all available in all nominal diameters and weights.

Each letter is characteristic of a chemical composition and sometimes implies a thermal treatment. The steels are produced in an open-hearth electric furnace or in a basic oxygen converter and all contain a maximum of 0.04% sulfur and 0.06% phosphorus. The number following the letter designates the minimum body yield stress guaranteed by the manufacturer and is expressed in thousands of psi: 80, for example, corresponds to a minimum body yield stress of 80 000 psi (approximately 550 MPa). The mechanical properties of the pipe can be deduced from this and from the geometrical characteristics.

Grades C75, L80 and C90, termed "special corrosion", are more specifically designed for wells with low temperature and low hydrogen sulfide concentration. They are characterized by a maximum body yield stress that is fairly close to the minimum (see Table 3.2).

B. Improved grades of steel (proprietary grades)

To meet the needs for hydrogen sulfide resistance, vendors offer low-alloy steels whose properties (superior to API standards) comply with NACE specifications. By way of an

Table 3.2 Grades and mechanical properties of tubing pipe (*Source:* After API Std. 5 CT, March 15, 1988).

Properties \ Grade	H40	J55	C75[1]	L80[1]	N80	C90[1]	P105
Color band identification[2]	1 black	1 green	1 blue	1 red + 1 brown	1 red	1 purple	1 white
Minimum yield stress (MPa)	276	379	517	552	552	620	724
(psi)	40 000	55 000	75 000	80 000	80 000	90 000	105 000
Maximum yield stress (MPa)	552	552	620	655	758	724	930
(psi)	80 000	80 000	90 000	95 000	110 000	105 000	135 000
Minimum tensile stress (MPa)	414	517	655	655	689	689	827
(psi)	60 000	75 000	95 000	95 000	100 000	100 000	120 000

1. Special corrosion.
2. Special clearance couplings (smaller diameter) must have a black line at the center of the color band.

example, Vallourec stocks a grade called L80 VH with a Rockwell hardness of 22. Note that an HRC (Hardness Rockwell C) of 22 is usually the maximum hardness authorized for standard steel to be considered H_2S service.

C. Stainless steel, alloys and special pipe

When wells containing carbon dioxide were put on stream the oil industry started using stainless steels. Some examples are 13% chromium steel if there is no H_2S in addition to the CO_2, and duplex 22-25% chromium steel when the hydrogen sulfide partial pressure is not too high, say a few tens of kPa (some psi). For even more severe duty, mainly high nickel content alloys are used, associated with chromium and molybdenum (Hastelloy). These special steels can cost up to twenty-five times more than standard steel.

3.3.1.4 Connections, threads

There are two ways to screw pipe together:
- by using a coupling, the most common connection, or
- by means of an integral joint, the most common type of connection on small diameter pipe (Fig. 3.5a).

Depending on the type of connection, the seal is separate from the thread or integrated in it.

A. API tapered triangular threaded connections

API proposes:

- NU (Non-Upset) coupling, with no increase in diameter at the end of the pipe body (Fig. 3.5b)
- EU (External Upset) coupling, with an increased diameter on the outside of the pipe body giving better tensile strength, as the cross-section at the base of the thread is larger (Fig. 3.5c).

API threads have a taper of 6.25% and 8 to 10 threads per inch. They are triangular and rounded on top. The seal is made by grease trapped between the threads, which is sufficient for low-pressure oil wells.

B. Premium joints with separate seals

Here the seal is made by a metal to metal shoulder or sometimes by an elastomer or Teflon joint (Fig. 3.5d). In the metal to metal shoulder connections, let us mention the following:

- Vallourec's VAM joint (Fig. 3.5e), commonly used by French companies, or Mannesmann's TDS joint, both with a modified tapered buttress thread
- Hydril's CS joint (Fig. 3.5f) with cylindrical trapezoid-shaped thread.

C. Greasing the threads

Grease is used to lubricate the threads and protect them from corrosion. For API-type threads, it also provides the seal for the connection. The couplings are screwed on in the factory with a friction factor one grease (except when otherwise specified). It is strongly recommended to use only greases with the same coefficient on the site so that the pipe makeup torque does not have to be modified. In this way, risks of under- or overtightening the connection can be avoided.

3.3.1.5 Mechanical characteristics of tubing pipe

The pipe run into the well is subjected to variable pressures and temperatures in addition to its own weight, and these will create variations in stress.

The main parameters are:

- body yield strength, expressed in newtons, tons or pounds
- collapse pressure, expressed in MPa or psi and based on the nominal thickness of the pipe
- burst, or internal yield pressure, also expressed in MPa or psi, but based on the minimum thickness authorized by the standard (87.5% of the nominal thickness).

Note that the mechanical characteristics are deduced from the nominal diameter and weight, and from the grade. For a 3 1/2" tubing, 12.7 #, N80, the cross-section of the pipe is 2375 mm^2 (3.68 in^2) and the thickness 9.52 mm (0.375 in). The body yield strength is 1.31 MN (294 klbf), the collapse pressure is 105.5 MPa (15300 psi) and the bursting pressure is 103.4 MPa (15000 psi).

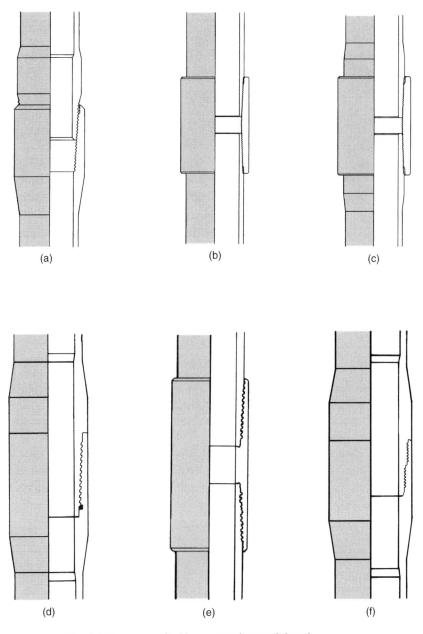

Fig. 3.5 Examples of tubing connections and threads
(*Source:* After *World Oil*, January, 1989).
a. API integral joint. b. API non-upset. c. API external upset.
d. Elastomer joint. e. VAM joint. f. CS Hydril joint.

The strength characteristics that must be taken into consideration are for the weakest part of the pipe (pipe body or connection, depending on the type of connection) depending on the type of connection. Using the same example as above, for the same pipe body, the body yield strength is only 1.1 MN (247 klbf) for the API NU connection. Meanwhile, it is the same or higher than that of the pipe body, i.e. 1.31 MN (294 klbf), for an API EU or a standard VAM connection.

3.3.2 Choosing the tubing

It may be useful to point out here that the choice of the tubing diameter has an obvious impact on the drilling and casing program and vice versa. In a given borehole, the tubing diameter that can be set in the production casing depends not only on the space available but also on the accessories that are to be incorporated into the tubing string.

The flow rate per well is the key parameter. It governs the number of wells that need to be drilled to achieve economic output of the field, hence the investments. This in turn conditions the profitability of the proposed development scheme.

The first parameter that needs to be chosen is the nominal tubing diameter. Then the grade of steel and the nominal weight are chosen based on the stress the tubing will have to withstand during production. Thirdly, depending on the type and corrosivity of existing and future effluents, the type of connection and the metallurgy are selected. In fact the different stages mentioned above overlap and sometimes make the choice of tubing a difficult job.

3.3.2.1 Determining the nominal pipe diameter

The nominal diameter via the nominal weight governs the inside through diameter of the pipe. The flows that can pass through it depend on the acceptable pressure losses of course but are also limited by two parameters: the maximum flow rate corresponding to the erosion velocity and the minimum flow rate necessary to achieve lifting of water or condensate by the gas. Note also that the space required by couplings of the tubing (or tubings in the event of a dual parallel string completion) limits the nominal tubing diameter that can be run into the production casing.

As an indication:

- API recommends keeping below a critical speed of approximately:

$$V_{0(m/s)} = \frac{120 \text{ to } 150}{\sqrt{\rho \text{ (kg/m}^3\text{)}}} \qquad V_{0(ft/s)} = \frac{35 \text{ to } 45}{\sqrt{\text{density (ppg)}}}$$

which corresponds to:

- 4 m/s (13 ft/s) for an oil whose specific gravity is 0.85 (850 kg/m^3 or 7.71 ppg)
- 10 m/s for a gas whose density is 150 kg/m^3 (1.25 ppg) (in fact when the gas is not aggressive this figure is too conservative and is revised upward by operators who reckon on 20 m/s (66 ft/s) for clean, sand-free gases that are only slightly or not at all corrosive)

3. THE EQUIPMENT OF NATURALLY FLOWING WELLS

- In wells where liquid drops are carried up to the surface, the minimum velocity of the gas is approximately 2 m/s for a pressure of 10 MPa (or 6.5 ft/s for 1500 psi), 1.5 m/s for 15 MPa (or 4.2 ft/s for 2200 psi) and 1 m/s for 30 MPa (or 3.2 ft/s for 4500 psi).

Table 3.3 gives an idea of the flow rates that can be contemplated in tubings in the most popular sizes and nominal weights. These figures are put forward for the following conditions:

- oil: a pressure loss not exceeding 250 Pa/m (i.e. 10 psi per 1000 ft) and an average velocity of under 2 m/s (6.5 ft/s)
- gas: a pressure loss not exceeding 1000 Pa/m (i.e. 40 psi per 1000 ft) and an average velocity of under 10 m/s (33 ft/s).

Note that the Reynolds number calculated with the data from the table varies:

- from 10^4 to 10^5 for oil, which suggests partially turbulent flow usually with a laminar film against the wall
- from $2 \cdot 10^6$ to $2 \cdot 10^7$ for gas, which indicates a perfectly turbulent flow.

Table 3.3 Tubing diameters and potential flow rates (*Source:* ENSPM Formation Industrie).

Oil: Specific gravity = 0.85, kinematic viscosity $\nu = 5$ mm²/s (cSt).
  Criteria: $\Delta P_{friction} \leq 0.25$ MPa/1000 m (10 psi/1000 ft) and velocity $V \leq 2$ m/s (6.5 ft/s).
Gas: Specific gravity = 0.7, dynamic viscosity $\mu = 0.015$ mPa·s (cP);
  density $\rho = 150$ kg/m³ (1.25 ppg); pressure $P = 15$ MPa (2200 psi);
  compressibility factor $Z = 0.9$; clean, sand-free gas.
  Criteria: $\Delta P_{friction} \leq 1$ MPa/1000 m and velocity $V \leq 10$ m/s (33 ft/s).

Nominal tubing diameter (in)	Nominal weight[1] (lb/ft)	Inside diameter (mm)	Inside diameter (in)	Drift (mm)	Drift (in)	Oil flow rate (m³/d)	Oil flow rate (bbl/d)	Gas flow rate (for $P = 15$ MPa ≈ 2200 psi)[2] (10^3 Sm³/d)	Gas flow rate (10^6 ft³/d)
2 3/8"	4.6	50.7	1.945	48.3	1.901	150	900	150	5
2 7/8"	6.4	62	2.441	59.6	2.347	275	1 700	275	10
3 1/2"	9.2	76	2.992	72.8	2.867	450	2 800	450	16
4"	10.9	88.3	3.476	85.1	3.351	700	4 400	700	25
4 1/2"	12.6	100.5	3.958	97.4	3.833	1 000	6 300	1 000	35
5 1/2"[3]	17	124.3	4.892	121.1	4.767	1 700	11 000	1 700	60
7"[3]	29	157.1	6.184	153.9	6.059	3 000	19 000	3 000[4]	105[4]
9 5/8"[3]	47	220.5	8.681	216.5	8.525	7 000[4]	44 000[4]	6 000[4]	210[4]

1. Most common nominal weights, VAM threaded pipe.
2. For higher pressures and densities, the possible gas flow rates under standard conditions can be greater (and vice versa if they are lower).
3. Casing pipe used as tubing.
4. Flow rate is not limited by the pressure loss criterion but by the maximum velocity criterion, such that velocity $V \leq 2$ m/s (6.5 ft/s) for oil or $V \leq 10$ m/s (33 ft/s) for gas.

3.3.2.2 Determining the grade and the nominal weight

Depending on the depth (and independently of the nominal weight) a minimum grade of steel is required to withstand the tensile stress due to the weight of the tubing itself. Moreover, the combination of grade and nominal weight determines pipe's internal yield pressure and collapse resistance.

During its productive life, the well experiences variations in flow rate, pressure and temperature which induce mechanical stress in the pipe, or if the tubing is free to slide in the packer or is equipped with a slip joint, pipe stretching or shortening, i.e. tubing "breathing". Using Lubinski's method, the tensional and compressional stresses, or elongation and contraction, will be determined. These stresses can be expected in the **adverse** situations that occur during the lifetime of a well, usually: tubing pressure testing, production, well shut-in, pay zone acidizing or even fracturing. The parameters that are assessed in this way are compared to the maximum authorized values by assigning safety coefficients to each of the original mechanical characteristics for pipe. In severe cases the method of compound stresses is increasingly used. For example, the ellipse of biaxial yield stress shows the decrease in collapse strength when considerable tensional stress is applied.

For a given tubing and completion configuration, the tubing should be checked to be sure it does not undergo:

- too much tensional stress at its uppermost end
- a tensile force at the packer which is greater than the tension required to unseat it
- excessive buckling or permanent "corkscrew" deformation
- elongation and contraction of excessive amplitude if movement is possible
- overly high bursting stress at its uppermost end
- overly high collapse stress in the lower part of the annulus.

These considerations are used to be sure whether the grade-nominal weight combination that was initially chosen is suitable or whether another choice should be envisaged. In practice it is less costly to increase the nominal weight (though some inside diameter is lost) than to choose a higher grade of steel and, in addition, this improves the H_2S resistance. Likewise, these calculations may suggest using a slip joint in some cases in order to deal with expansion and contraction in the production string.

3.3.2.3 Determining the connection and the metallurgy

A. Connections, type of thread

API connections are perfectly well adapted and sufficient for most oil wells. In contrast, joints with a sealing shoulder separate from the thread, usually metal to metal, are preferred for gas wells. Additionally, high pressure but also a corrosive effluent suggest the use of premium joints. In these instances French companies tend to recommend metal to metal connections of the VAM or equivalent type.

B. Metallurgy

To meet corrosion resistance needs, pipe can be made of normal or slightly treated steel, with the injection of a corrosion inhibitor which generates a protective film on the pipe wall, or it can be manufactured of alloy steel. The choice depends on:

- the cost per unit weight of the alloy steel
- the feasibility and cost of injecting corrosion inhibitors
- the time frame scheduled between two workover operations, their cost and duration as well as the cost of "lost" production during these operations.

In sensitive cases the tendency is to prefer an alloy steel to inhibitor injection.

It can also be pointed out that:

- Choosing an inhibitor, determining the right dosage, and implementing inhibition are operations that require specialists. In addition, the products used can unfortunately cause problems downstream when the oil and/or gas are treated.
- An alloy steel, designed and manufactured to withstand a certain corrosive environment can, however, become quickly corroded in another environment, particularly in the open air and especially in a marine air environment. This means very strict storage and transportation conditions. Moreover, the use of alloy steels requires the problem of corrosion by electrolysis (also termed electrolytic or galvanic corrosion) to be solved. When alloy steel elements are in contact with others of a different metallurgy, made of more conventional carbon steel for example, one of these steels will behave like a soluble anode.
- Pipe made of alloy steels, especially the stainless steel pipe widely used in carbon dioxide environments, can cause seizure problems and requires special care during tightening-loosening operations. However, pipe manufacturers have made a lot of progress recently in special threads and coatings and the seizure problem is now less serious.

3.4 PACKERS

Packers are usually run in and set in the production casing or the liner. They protect the annulus from corrosion by formation fluids and limit the pressure in the annulus so that the casing and its cement sheath do not undergo overly wide variations in compressional stress. The presence of a packer makes it possible to put a fluid, called packer or annular fluid, in the annulus which also helps protect the casing. In short, the packer isolates the annulus from any physical contact with the formation fluids and bottomhole pressure.

3.4.1 Packer fluids (or annular fluids)

Placed in the annulus, packer fluids serve mainly to protect the casing. They also allow the differential pressure on either side of the packer to be reduced, thereby limiting hydraulic stresses at packer level. They help counteract the casing's collapse stress and the tubing's bursting stress in the lower sections of the pipe.

Note also that a packer fluid can help control a well if there is a leak in the production tubing or when the packer no longer provides a seal or has been deliberately unseated.

Fluids containing suspended solids are not used as packer fluids as the solids settle out and may make workover operations more difficult. In practice depending on the required specific gravity, brine, diesel oil or water are commonly used. In addition, these fluids are treated with a corrosion inhibitor, a bactericide and an antioxidant.

3.4.2 The main packer types

A packer is basically defined by the setting mechanism, the seal, the means of retrieval and the type of tubing-packer connection. They are usually classified based on the means of retrieval as the first criterion.

- Packers are set by steel slips which, when pushed along a cone-shaped ramp, grip the casing. A seal is provided by rubber rings squeezed out against the casing.
- Packers can be retrieved in three different ways:
  - Simply drilling out or milling the packer: this is the case for permanent production packers.
  - Actuating shear pins or rings by pulling on the tubing. This movement releases the anchoring slips. This is the case for retrievable packers. Some of them require a special retrieving tool.
  - Mechanically releasing without actuating shear pins or rings: this is the case for "temporary" mechanical packers that are mainly used in special strings for well testing, remedial cementing, acidizing, etc.
- The tubing-packer connection can be of two types:
  - Rigid: the tubing is fixed onto the packer.
  - Semi-free: the tubing enters the packer by means of a pipe with sealing elements which can slide freely up and down. This system allows the tubing to move up and down (variations in tubing length due to "breathing"). The allowable sliding range depends on the length of the pipe equipped with sealing elements and on its initial position. Additionally, the downward movement is limited by a stop guard.

3.4.3 Choosing the packer

The choice of packer type depends on:
- the resistance of the packer and setting mechanism to mechanical and hydraulic stresses in the well:
  - allowable differential pressure

- allowable compression and tension at the tubing-packer connection and casing-packer contact,
- maximum temperature for the elastomers
- setting and retrieval procedures
- available accessories
- consequences and costs in initial completion and workover operations
- the packer's reputation and the user's experience with it.

Further considerations are as follows when a packer is being chosen:

- the inside diameter of the casing
- the packer's inside through diameter
- elastomers' resistance to fluids
- metallurgy (corrosion problems).

3.4.4 Permanent production packers

Permanent packers are also called drillable packers. The typical example of this sort of packer is Baker's 415 D permanent packer (Fig. 3.6).

3.4.4.1 Description of the 415 D packer

The packer consists of an inner mandrel (1a), ground and polished inside, surmounted by a left-hand square thread (1b).

Outside the inner mandrel from bottom to top are:

- a bottom guide (2) screwed on the inner mandrel
- lower anchoring slips (3) and their anchoring cone (4)
- a rubber packing element (6) with lead anti-extrusion rings (5) and (7)
- upper anchoring slips (9) and their anchoring cone (8)
- an outer setting sleeve (10) cotterpinned to the inner mandrel (1a).

A ring ratchet (11) allows the outer setting sleeve (10) to slide down in relation to the inner mandrel (1a) but prevents the reverse, thereby making the setting operation permanent.

This packer is entirely drillable, all the parts except for the packing element are made of cast iron, magnesium, lead or bronze.

It is run and set prior to running the final tubing either on the electric cable equipped with an appropriate setting tool, or on a drill or tubing string with the corresponding setting tool.

3.4.4.2 Setting the 415 D packer on the electric cable

In a straight or moderately deviated well, the most common way of setting the 415 D with its under-packer extension (provided the total weight does not exceed the cable's capacity) is on

3. THE EQUIPMENT OF NATURALLY FLOWING WELLS

Fig. 3.6 Permanent production packer (*Source:* After Baker catalog, 1984–1985).

the electric cable. It allows the packer to be precision-set in the well according to reservoir data. In addition, it is a quick, practical and safe solution.

The setting tool is connected to the electric cable, usually the same one used by the logging company. Schematically, this tool is a hydraulic jack that uses the energy produced by expanding gases when an explosive charge is fired.

The tool includes (Fig. 3.7):

- a cylindrical body (2) attached at its top to the cable and at its bottom to the square thread (1b) of the packer inner mandrel (1a) by means of a segmented retractable thread (3)
- a piston (4) fixing bar (5) adapter coupling (6) assembly that rests against the setting sleeve of the packer (10).

When the packer reaches the required depth and after depth matching, the explosive charge (7) placed in the upper chamber (8) of the tool is fired electrically. The force of the expanding gases is transmitted to the hydraulic jack via a floating piston (9) which expels the oil contained in the upper cylinder (11) toward the lower cylinder through a flow bean (12).

After the bronze pins that keep the setting cones from moving while running are sheared, the inner mandrel (1a) and the setting sleeve (10) move in relation to each other, compressing the rubber packing element and forcing the slips to move outward as they slide up the cones. The slips grip the steel and become embedded in the casing.

At the end of the setting operation, part of the hydraulic energy generated by the explosive is used to release the setting tool. The final pull exerted by the jack shears a release stud (13) and releases a small mandrel (14) allowing the divided thread to retract by elasticity. The tool is then pulled out of the hole.

3.4.4.3 Setting the 415 D packer on a pipe string (Fig. 3.8)

The setting tool in this case consists of a hydraulic jack. The piston is screwed into the left-hand square thread (1b) at the top of the inner mandrel (1a), the body of the jack rests against the setting sleeve (10). The jack is locked when it is run in by means of shear pins.

At the required depth a ball is dropped from the surface and comes to rest on its seat in the setting tool. Pumping through the pipe string with the drilling pumps presses the ball against its seat, thereby increasing the pressure. At the required value, the pressure in the jack is sufficient to shear the pins. The packer can then be set. The jack is used to set the upper slips, then pulling on the pipe helps compress the packing element and set the lower slips.

After the pressure in the pipe string is bled off, the proper setting of the packer is checked by means of a tension-compression test. The sealing capacity of the packing element is also tested by moderately pressurizing the annulus.

The setting tool used with the pipe string is released by rotating to the right (10 rotations on the bottom) and the string is then pulled out of the hole in order to remove the setting tool.

Note: There are now other permanent packers along the lines of the 415 D. They can be set hydraulically directly with the final production string run into the well at the same time. There is no longer any setting tool, and setting is hydraulic, similar to the method used for retrievable hydraulic packers (see Section 3.4.5).

3.4.4.4 Connection between the tubing and the 415 D packer

There is a choice between the two connections described below.

A. *Anchor seal assembly* (Fig. 3.9a)

An elastic thread fixes the tubing to the packer. A seal is provided by sealing elements. The assembly should comprise at least two seal stacks.

B. *Locator seal assembly* (Fig. 3.9b)

This equipment has got only sealing elements and allows the tubing to slide in the packer. However, a stop guard located in place of the thread limits the downward stroke.

3. THE EQUIPMENT OF NATURALLY FLOWING WELLS

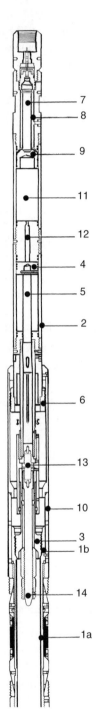

Fig. 3.7 Electric cable setting tool for the Baker permanent packer (*Source:* After Baker catalog, 1984–1985).

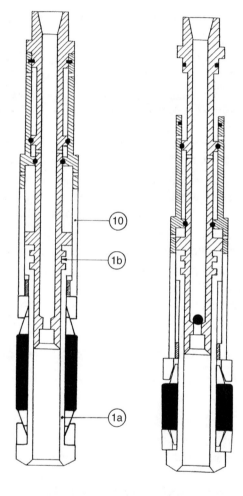

Fig. 3.8 Simplified diagram of the permanent packer setting tool that is run in with a pipe string (*Source:* ENSPM Formation Industrie).

3. THE EQUIPMENT OF NATURALLY FLOWING WELLS

Fig. 3.9a Anchor seal assembly. Type E Baker (*Source:* Baker catalog, 1984–1985).

Fig. 3.9b Locator seal assembly. Type G Baker (*Source:* Baker catalog, 1984–1985).

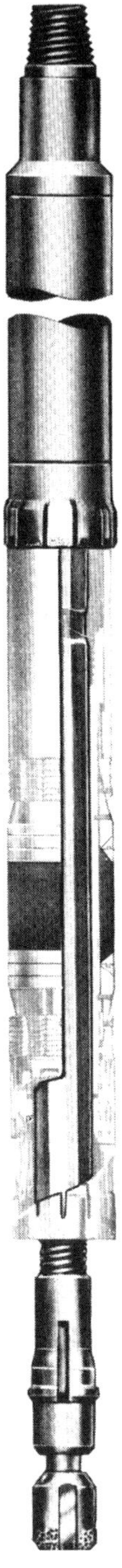

Fig. 3.10 Typical packer washover mill for a permanent packer (*Source:* Baker catalog, 1984–1985).

The length of the locator and its initial entry length inside the packer are chosen according to the amplitude of the movement caused by the tubing breathing during production. Additionally, it is strongly recommended to add a sufficiently long seal bore extension on under the packer. This is so that only the lower set of sealing elements is in contact with the oil and/or gas, whatever the entry length of the locator. This will extend the lifetime of the other sets of sealing elements.

3.4.4.5 Drilling out the 415 D packer

Once the tubing has been pulled out, the packer is drilled out in a few hours with a drill string and a full-gage reamer or a washover mill. When the upper part of the packer and the upper slips have been drilled out, the rest of the packer is automatically unseated and can be released downward. It can be pushed down and left in the bottom of the well.

If no extraneous material must be left in the well, the washover mill is used in conjunction with an inner mandrel that extends through the packer (Fig. 3.10). At the lower end of the extension there is a catch sleeve to retrieve the lower part of the packer after the upper part has been milled out. The tool features a release device used to free the drill string if the lower part of the packer gets stuck and can not be pulled out.

So that there is enough space under the packer for the tool extension when the inside diameter of the tubing under the packer is smaller than that of the packer, a mill-out extension of appropriate diameter is placed between the packer and the tubing.

3.4.4.6 Advantages and drawbacks of permanent packers

This type of packer is simple in design and does not include complicated mechanisms. It is highly reliable and can withstand considerable mechanical strains as well as high differential pressures. It also has a range that offers the largest inside through diameter to fluid flow for a given casing diameter. It is flexible as to the possible tubing-packer connections and can be left in the well during workover operations to change production equipment.

The biggest drawback is that it can only be removed by milling or drilling out. This means that a drill string-type assembly must be used after the production wellhead has been removed and the tubing pulled out. Another point that should be emphasized is that the production tubing sealing elements eventually adhere to the packer's inner mandrel if the tubing does not move for a long period. In contrast, overly frequent movement causes premature erosion of the sealing elements.

In any case, this type of packer has no equal in deep wells and is also often used in gas-producing wells.

3.4.5 Retrievable packers

These packers are designed to be unseated and pulled out of the well simply without having to be milled out. Therefore they all have an incorporated mechanism so they can be unseated.

Depending on the model, they are hydraulically or mechanically set and all are connected to the tubing permanently. However, a disconnection joint (or divider) can be included in the tubing string above the packer.

3.4.5.1 Hydraulically set retrievable packers (Fig. 3.11)

These packers are set by pressurizing the production string. The setting slips (1) are located under the seal (2), which usually consists of three rubber packing elements. The packing elements are often of different hardness and are chosen according to setting conditions and depth. They are separated by rings (3) that limit extrusion of the rubber. The slips keep the packer in place and prevent it from slipping downward as long as there is some "weight" on it. The use of hold down buttons (4) (friction buttons that are actuated hydraulically when the pressure under the packer is higher than the annular pressure) helps keep the packer from slipping upward.

A. Setting hydraulic packers

After a plug is set in a landing nipple under the packer or a ball is dropped and comes to rest on an appropriate seat, the tubing is pressured up by pumping until the packer's shear pins are broken. The pressure then acts on a piston inside the packer which drives a sleeve. This makes a tapered ramp move down expanding and anchoring the slips in the casing wall, while compressing the rubber packing elements that bulge out and provide the seal against the casing. In the same way as permanent packers, an inside ratchet system keeps the main slips in the set position.

B. Unseating retrievable hydraulic packers

These packers are unseated by pulling on the tubing until tensional strain is sufficient to shear the pins or ring (5) which are calibrated versus a tension on the tubing that is greater than any tension that may be reached during production.

After the pins or ring are sheared, the relative movement of the parts in the packer opens a pressure equalization passage allowing the hold-down buttons to withdraw, the packing elements to be released and the slips to retract.

The packers can not be reused immediately, but must be reconditioned first. In particular, the shear pins or ring must be replaced and the packing elements and seals must be checked.

C. Advantages and limitations of retrievable hydraulic packers

These packers are more complicated than permanent ones, but have practically no equivalent when the aim is to set a packer directly with the final production tubing. They are relatively easy to run in deviated wells, they are available in a dual and even triple version for multiple string completions. It is also possible to set several at the same time in the well. Finally, it should be noted that they are theoretically easy to unseat and retrieve. Once they have been pulled out, they can be reconditioned and reused.

3. THE EQUIPMENT OF NATURALLY FLOWING WELLS

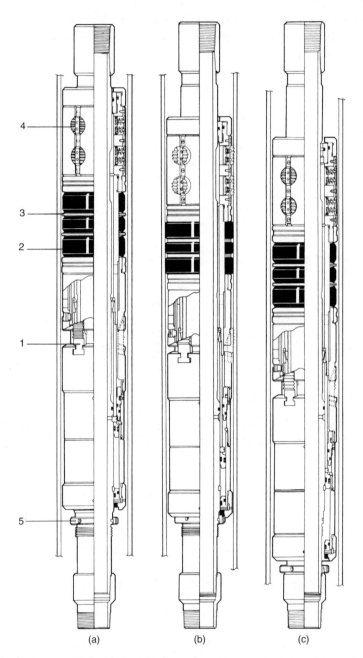

Fig. 3.11 Baker FH hydraulic packer (*Source:* After Baker catalog, 1984–1985).
a. Running in. b. Setting. c. Unseating.

3. THE EQUIPMENT OF NATURALLY FLOWING WELLS

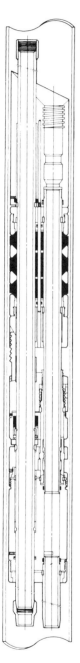

Fig. 3.12 Dual production packer. Baker's model A5 (*Source:* Baker catalog, 1984–1985).

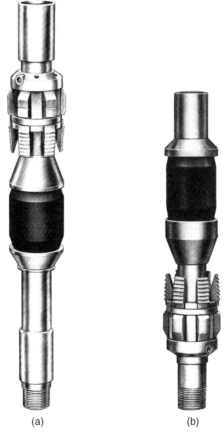

(a)   (b)

Fig. 3.13 Mechanically set retrievable packers (*Source:* Baker catalog, 1984–1985).
a. Tension setting. b. Compression setting.

However, the integrated hydraulic setting and unseating system is expensive in terms of space and accordingly limits the through diameter available for the fluid in a given casing. Setting requires a plug and a landing nipple under the packer or a ball dropped on a retractable seat. The plug may get stuck or the ball seat may refuse to retract. Lastly, it is not designed or manufactured of materials that are readily milled. Consequently, drilling out a packer of this type that has gotten stuck in the well can be extremely costly in terms of time, money and tools.

Retrievable hydraulic packers are available in the dual version (Fig. 3.12). The setting and retrieving procedures are similar to those described above. Depending on the model chosen, they are set by increasing pressure in the short string or in the long string. The packers are unseated by pulling on the long string.

3.4.5.2 Mechanically set retrievable packers (Fig. 3.13)

These packers are seldom used in permanent production strings. In contrast they are perfect for temporary stimulation, cementing and testing strings. This is because they can be reset immediately without having to be pulled out and because they are simple to retrieve. Generally speaking, these packers are set by compression, tension or rotating the string. They are equipped with friction pads to release and actuate the slips by rotating 90° in a J slot. Unseating them is very easy, usually by using the reverse of the setting procedure.

3.5 DOWNHOLE EQUIPMENT

Certain particular pieces of equipment are set below and above the packer. They are chosen because they are useful during completion, operations on the well and well control. They often involve using wireline techniques.

3.5.1 Circulating devices

The device is placed above the packer and allows the tubing and the annulus to communicate if need be. This proves very useful when equipment is set in the well and the well is brought on stream. It can also be used to control the well during preliminary workover operations. Choosing the circulating system from among the tools offered by manufacturers is mainly a question of reliability, ease of operation, compatibility with other special pieces of well equipment and the possibility of in situ repair or reconditioning with only a small-scale operation on the well, especially by wireline.

It should not be forgotten that any circulating system between the tubing and the annulus is a potential source of leakage or sticking. In extreme cases, this requires all the well equipment, or sometimes only the part up to the tubing-packer connection, to be pulled out in order to replace it.

3.5.1.1 Sliding sleeve circulating valve (Fig. 3.14a)

This type of circulating device is very common and is better known under the name sliding side door or sliding sleeve, SSD or SS. The tubing and annulus communication is established or cut off by means of a sliding sleeve (1).

By means of a shifting tool lowered on the slick line, the sleeve (1) is shifted so as to open or close the ports (2) machined in the body of the valve (3) (see Section 5.2.1 about the wireline string and the required surface equipment). Special machining of the body allows "spring" fingers (4) to lock the sleeve in the required position.

The cross sectional area of the ports is larger than the cross-section of the tubing and as a result, relatively high fluid flow rates can be contemplated, without running the risk of systematically eroding the valve. In contrast, the seals (5) and (6) must be expected to deteriorate after the sliding sleeve has been shifted a certain number of times. This causes leakage especially in the presence of gases or sediments. Accordingly, it is absolutely not recommended to operate the sliding sleeve except for initial completion or workover.

3.5.1.2 Side pocket mandrel (Fig. 3.14b)

This mandrel is normally designed for artificial lift by gas lift (see Chapter 4), but it is used on a number of flowing wells. The side pocket is equipped with a dummy instead of a gas-lift valve. When circulation is required, the dummy is fished out by wireline and replaced by a simple ported sleeve, designed to protect the seal bores in the side pocket.

The mandrels are also used to inject chemicals into the tubing through the annulus: demulsifying agents and corrosion inhibitors in particular. This can be done by equipping the mandrels with injection valves that are opened by a specified value of annulus overpressure.

The main advantage of such a system is that the seals that surround the communicating ports are carried by the tool placed in the side pocket and are consequently easy to replace. Also appreciated is the fact that the side pocket mandrel causes no diameter restrictions in the tubing for production or running in other tools by wireline, snubbing or coiled tubing (see Chapter 5). The major drawbacks are first of all a small cross-sectional area for communication with the annulus (designed for gas), which does not allow very high liquid flow rates. Furthermore, substantial additional outside space is required in relation to the tubing diameter (in fact mandrels are not available on the market for large tubing diameters), which is not automatically compatible with a casing that is technically or economically acceptable.

3. THE EQUIPMENT OF NATURALLY FLOWING WELLS

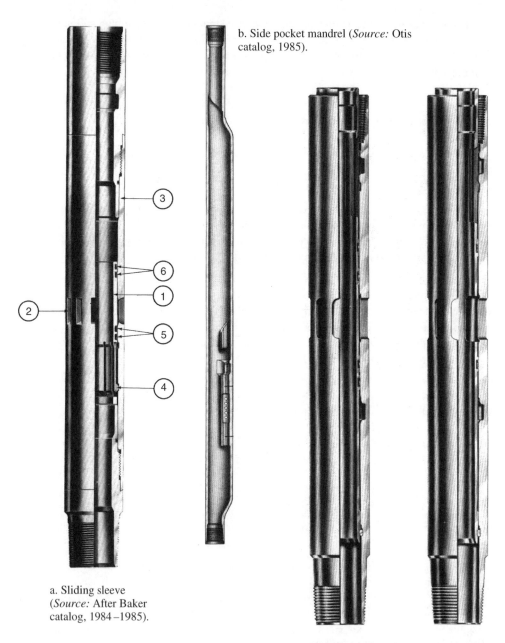

a. Sliding sleeve (*Source:* After Baker catalog, 1984–1985).

b. Side pocket mandrel (*Source:* Otis catalog, 1985).

c. Ported landing nipple (*Source:* AVA catalog).

Fig. 3.14 Circulating devices.

3.5.1.3 Ported landing nipple (Fig. 3.14c)

This is a specific type of landing nipple (see Section 5.2) that features ports. Communication is established or prohibited by a blinding sleeve equipped with seals that is run or retrieved by wireline. The sleeve (and therefore the seals) is very easy to pull out and replace in the event of leakage. In contrast, the through diameter restriction is rather unfavorable if other wireline operations have to be carried out deeper.

3.5.1.4 Conclusion, comparison

The circulating device that offers the largest cross sectional area is undoubtedly the sliding sleeve valve, which is consequently installed in many wells. The side pocket mandrel allows the part carrying the seals to be retrieved and, because of this, is certainly advantageous for wells where workover is costly or fairly complex, such as subsea wells.

Note that as a last resort, particularly to get the well under control at the beginning of a workover operation, the tubing and annulus can be made to communicate, allowing the fluids to circulate in the well, by making a hole in the tubing with a perforator run in on a wireline.

Lastly, let us point out that whatever the circulating system used, care must be taken when it is opened. To prevent the wireline assembly from being ejected or subjected to an overly high tensional stress, the tubing and annulus pressures opposite the device must be (approximately) equalized.

3.5.2 Landing nipples

In order to meet measurement requirements, facilitate equipment installation or perform other safety functions, the tubing is equipped with special pieces of equipment, called landing nipples, where mechanical tools can be seated, usually by wireline.

The basic requirements are to:

- pressure test all or part of the tubing
- test the tightness of the circulating device
- pressure up the tubing to set a hydraulic packer
- isolate the tubing from formation pressure
- leave pressure and/or temperature measurement instruments in the well temporarily, if possible without interfering too much with the well's production conditions.

A number of tools are available to fulfill any one of these functions. They are usually screwed on under a mandrel (or a lock mandrel) that keeps them in place in the landing nipple.

During completion, the location, number and type of landing nipples in the well must be given very careful consideration and chosen according to:

- any operations that may be performed in the well at a later date
- the loss of inside through diameter due to the landing nipple, especially for the tools that have to be run deeper into the well.

Generally speaking, it is advisable to limit the number of landing nipples to a strict minimum, in most cases two or three will be enough.

Several types of landing nipples exist on the market, but they all have at least two points in common:

- A locking groove allowing the tool to be mechanically locked in the landing nipple if need be, in which case the mandrel is equipped with a lock.
- A seal bore where the seal is made between the landing nipple and the tool if need be by means of V packing type seals mounted on the mandrel. It should be noted that the diameter of this seal bore serves as a reference diameter for the landing nipple under the term nominal diameter, expressed in decimals limited to hundredths of an inch (for example 2.81" for 2.812" in reality). In order to be able to set the corresponding tool in this landing nipple without the tool getting stuck in the tubing while it is being run in and without the seals being overly damaged, the nominal diameter of the nipple must at least be smaller than the drift of the tubing.

Although manufacturers supply landing nipples whose nominal diameter is standardized with respect to tubing, nipple as well as lock profiles are, however, particular to each company. In other words, the type of nipple and its brand name often condition which tools can be set in it and vice versa unless adapters are used.

There are, however, two main categories of landing nipples that we will call "full bore" and "bottom no-go":

- Full bore landing nipples are characterized by a through diameter that is equal to the nominal diameter of the landing nipple, i.e. of its seal bore. It should be noted though, that these landing nipples cause a restriction with respect to the tubing inside diameter as such (remember that the nominal diameter of a landing nipple must be at least smaller than the drift of the tubing).

  In this category there are the following:
  - full bore simple, often called only full bore
  - full bore selective, often called only selective
  - full bore top no-go, often called only top no-go.
- Bottom no-go landing nipples are often called only no-go and are characterized by a through diameter that is smaller than the nominal diameter of the seal bore: a shoulder causes a restriction at the lower end.

Among the products available on the market, French companies commonly use those by Baker and Otis, and to a lesser degree those by AVA and Camco.

3.5.2.1 Full bore simple landing nipples (Fig. 3.15)

These landing nipples have only got a locking groove and a seal bore, they do not restrict fluid flow other than by their nominal diameter. As a result, as many of them as needed of the same nominal diameter can theoretically be placed in a well. In practice, the number is limited to four

3. THE EQUIPMENT OF NATURALLY FLOWING WELLS

landing nipples of the same nominal diameter in order to allow for possible damage to the mandrel packings when they pass through the landing nipples above the target landing nipple.

The equipment is run with a running tool that keeps the locking dogs on the mandrel retracted so as not to impede running (the same holds true for retrieving it). The equipment can therefore pass the other landing nipple(s) of the same nominal diameter located above the target and the mandrel can also be termed full bore by extension.

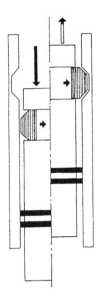

Fig. 3.15 Full bore simple landing nipple (*Source:* ENSPM Formation Industrie).

The landing nipple where the tool is to be placed is selected as follows: the assembly passes the landing nipple and is then pulled up. This actuates the locking dogs (or rather disables the dog retracting mechanism). Here it is the running tool and the running procedure that provide selectivity.

Manufacturers generally call these landing nipples full bore and sometimes even selective by mistake, since it is the setting procedure rather than the landing nipple that allows selection. Some examples are Otis's X and R and Camco's W landing nipples. They are also found incorporated in some circulating devices such as Otis SSD type XO, XA, RO, RA and Camco SSD type W.

In addition to limiting the number of landing nipples, it is advisable to space them out at a distance of a few tubing lengths so that there are no doubts about which landing nipple the wireline operator is aiming for.

Lastly, it should be pointed out that this is the selection method used for side pocket mandrels (see Chapter 4).

3.5.2.2 Full bore selective landing nipples (Fig. 3.16)

This type of landing nipple is offered by Otis under the designation S or T. In addition to the features of the preceding type of landing nipple, it has a selection profile. Seven different inside profiles (or "key profiles") are available and are indicated by a number of circles from one to seven engraved on the outside of the landing nipple. Selectivity is achieved by choosing the right selection key profile on the mandrel corresponding to the inside profile of the landing nipple where the tool is to fit (Fig. 3.16a).

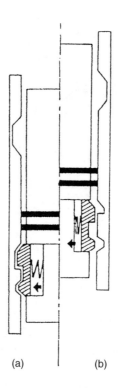

(a)   (b)

Fig. 3.16 Full bore selective landing nipple (*Source:* ENSPM Formation Industrie).

There is, however, an order to be followed in positioning the S-type landing nipples in the well. Given the way the profiles are machined, the landing nipple with the least number of circles must be set the lowest, then the others are set in order of increasing number of circles coming upward in the well. The progressively machined selection profile of the landing nipple (from one to seven "fingers" on the hand) means that key 1 can only fit into landing nipple 1 and consequently does not stop at the others (Fig. 3.16b). In contrast, key 2 can fit into landing nipples 2 and 1, key 3 into 3, 2 and 1 and so on. This shows how important it is to set the landing nipples in the right order from bottom to top.

3.5.2.3 Full bore top no-go landing nipples (Fig. 3.17)

The upper part of these landing nipples is oversize in comparison with the seal bore. This configuration allows a mandrel with a no-go ring of a diameter larger than that of the landing nipple's seal bore above its packing section to stop in this landing nipple. The mandrel is then termed top no-go and by extension the landing nipple is also often called top no-go.

Depending on the mandrel, the downward locking by the no-go can be used to release, by means of downward jarring, the locking dogs that lock the mandrel upward. Full bore mandrels can also be set in these landing nipples with a selective running tool.

Given the landing nipples' specific features, great care must be exercised when choosing them, especially if top no-go mandrels are used. It is absolutely necessary to check that the maximum diameter of the mandrel's top no-go ring is really smaller than the drift of the tubing it is to be run through. During the setting operation, the tool should be seated gently on its landing nipple, otherwise the top no-go ring may well get crimped in the landing nipple (the diagram is not to be trusted in this respect, as dimensions are not to scale).

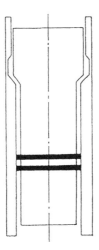

Fig. 3.17 Full bore top no-go landing nipple (*Source:* ENSPM Formation Industrie).

These landing nipples are well suited for accommodating pressure testing plugs and usually withstand 70 MPa (10 000 psi).

Depending on the manufacturer and the different possible methods of use as discussed above, these landing nipples are sometimes called top no-go, but also full bore or selective. Despite the name "selective", it is the running tool alone and not the landing nipple that is selective. Some examples are Baker's F, Camco's DB or D, AVA's SEL landing nipples. They can also be found incorporated in some circulating devices (SS model by Baker, for example).

3.5.2.4 Bottom no-go landing nipples (Fig. 3.18)

These landing nipples feature a machined in shoulder at the base of the seal bore. The shoulder is called a bottom no-go and the corresponding mandrel is stopped by it. The mandrel diameter is smaller than, but very close to, that of the seal bore. Because of this and so that there is good contact and the mandrel is prevented from sticking, there is an appropriate profile machined out below the packing section.

Because of the type of landing nipple it fits into, the mandrel is usually called bottom no-go or even sometimes non-selective. As for the top no-go mandrel, the bottom no-go can include a system of locking dogs that lock the mandrel upward.

To keep from "losing" too much diameter for equipment that is to be set below, the diameter of the polished bore and of the bottom no-go are close to each other (for example, Baker's R type 2.75" landing nipple: seal bore 69.85 mm [2.750"]; no-go 68.50 mm [2.697"]).

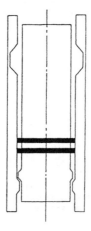

Fig. 3.18 Bottom no-go landing nipple (*Source:* ENSPM Formation Industrie).

During the setting operation the tool must also be seated gently on its landing nipple in the same way as for the full bore top no-go and even more so. Likewise, care needs to be taken during pressure testing when plugs are used with a bottom no-go as they have a nasty habit of getting crimped in the no-go (do not exceed 35 MPa or 5000 psi of differential pressure on the plug). The completion equipment may then have to be pulled out in order to retrieve them. Manufacturers usually call these landing nipples bottom no-go or sometimes simply no-go. Some examples are Baker's R, Otis's XN and AVA's BNG landing nipples.

3.5.2.5 Landing nipple summary table

Table 3.4 summarizes the main types of landing nipples supplied by manufacturers along with the relevant operating features for each of the two basic categories: full bore and bottom no-go.

Table 3.4 Landing nipples summary table (*Source:* ENSPM Formation Industrie).

CATEGORY		TYPE	MEANS OF SELECTION
FULL BORE	SIMPLE	**Full bore simple** Otis type X, R* Camco type W *also on Otis SSD type XO, XA, RO, RA	Running tool actuates locking dogs when **pulling upward** on mandrel Maximum mandrel diameter: < landing nipple nominal diameter
FULL BORE	SELECTIVE	**Full bore selective** Otis type S, T BE CAREFUL to install them in the right order: 7 ↑ 1	**Selection key** on the **mandrel** fits into the **selection** profile of the **landing nipple** Maximum mandrel diameter: < landing nipple nominal diameter
FULL BORE	TOP NO-GO	**Full bore top no-go** Baker type F** Camco type D, DB AVA type SEL ** also on Baker SS type L	a) **Top no-go ring** on **mandrel** is oversize compared to the landing nipple's nominal diameter Maximum mandrel diameter: > landing nipple nominal diameter *Be careful that:* *top no-go mandrel diameter:* *< tubing drift diameter* b) Or same as for full bore simple
BOTTOM NO-GO		**Bottom no-go** Baker type R Otis type N, XN Camco type DN AVA type BNG Landing nipple bottom no-go diameter: < landing nipple nominal diameter	**Landing nipple (and Mandrel)** with a **bottom no-go ring** Maximum mandrel diameter: < landing nipple nominal diameter > landing nipple bottom no-go diameter

3.5.2.6 Using several landing nipples in the same tubing

In the same completion several different types of landing nipples can be used together, however in practice they are not usually mixed up and the choice usually tends to be only:

- Full bore selective landing nipples with the same nominal bore differentiated by their selection profiles.
- Or no-go landing nipples.

  Note that tools designed to stop in a bottom no-go landing nipple can be run through all the full bore top no-go, full bore simple, and full bore selective landing nipples of the same nominal diameter. As a result, the following sequence can exist using for example Baker's F and R type landing nipples:

  - F 2.81 full bore top no-go [nominal landing nipple diameter: 71.42 mm (2.812 in), tool no-go ring diameter: 72.77 mm (2.865 in)]

  - F 2.75 full bore top no-go [nominal landing nipple diameter: 69.85 mm (2.750 in), tool no-go ring diameter: 71.17 mm (2.802 in)]

  - R 2.75 bottom no-go [nominal landing nipple diameter: 69.85 mm (2.750 in), bottom no-go landing nipple diameter: 68.50 mm (2.697 in), tool no-go ring diameter: 69.60 mm (2.740 in)] and so on.

- Or full bore simple landing nipples with the same nominal diameter, and selection by the running tool that allows locking only when pulling. It is advisable to space them out sufficiently along the tubing so that there is no possible doubt as to which landing nipple the wireline operator is working on.

In practice, bottom no-go landing nipples are not normally used in the tubing, except for the lowest nipple. This is true whatever the type of landing nipple located above it. Choosing a bottom no-go for the lowest landing nipple affords the following advantages, provided there is a suitable wireline string available:

- the wireline string will not mistakenly pass the lower part of the tubing, an asset since getting it back inside can be tricky, especially in a deviated well
- the wireline string can be stopped at this point if the cable fails and retrieving will be easier in the tubing (where the tool is centered) than in the casing at the bottom of the hole.

Note also that:

- In order to delay the time when the tubing is washed out from the inside due to turbulence generated by the severe restriction of the internal cross-sectional diameter caused by the presence of certain tools (bottomhole choke, SCSSV), it is possible to add reinforced thickness elements called flow couplings on either side of the landing nipples.
- Certain subsurface safety valves are set in special landing nipples that communicate with the surface by means of a special small hydraulic line located in the annulus.

- There are mandrels that are directly locked on the tubing wall, i.e. without any special landing nipple. They are, however, not used under normal conditions because there is always doubt as to the locking quality and because the differential pressure they can withstand from bottom to top is limited.
- When the equipment set is a plug, it is always equipped with a device that allows pressures to be equalized on either side before the mandrel is released. Equalization is not immediate, it requires a certain amount of time.

3.5.3 Other downhole equipment

a. Equipment below the packer

A perforated tube can be incorporated in the tubing string with a bottom no-go landing nipple at its lower end designed to accommodate the pressure and/or temperature recorder holder. This allows pressure measurements to be made during production without introducing further pressure losses due to the recorder restricting the cross-sectional area.

For multiple zone completions, reinforced thickness pipes, called blast joints, must be used opposite perforated zones to delay washing out from outside by fluids gushing out of the casing perforations.

Between several packers, for example in a parallel dual string completion or an alternate selective completion, and between packer and screens in wells with sand control, a safety joint is generally incorporated to facilitate selective pulling out of downhole equipment.

b. Equipment above the packer

The use of a slip joint allows variations in tubing length due to changing well conditions (temperature, pressure, flow rate) and prevents excessive extra strains on the packer and the tubing itself.

A disconnection joint can often be used as a slip joint and allows the tubing to be pulled out without unseating the packer. The well is kept under safe conditions by setting a plug in a landing nipple that is usually machined in the lower half-joint fixed to the packer and by pumping a control fluid into the well. If the plug is set before the control fluid is circulated, the fluid is prevented from coming into contact with the formation and there are fewer risks of losses or damage.

3.6 SUBSURFACE SAFETY VALVES

Depending on the environment and on the type and pressure of the produced effluent, it may be necessary to place a SubSurface Safety Valve, SSSV, inside the well itself. It supplements the one(s) on the wellhead if it/they should happen to be out of order (valve failure, wellhead torn off, etc.).

The first valves to have been used are known as storm chokes and are set in a landing nipple in the tubing by wireline, often near the packer. They are totally independent and close when the flow rate rises abnormally or when there is too much pressure drop across them. This is also their main drawback, because they can not be actuated when trouble is detected on the surface. They can not be closed deliberately unless the well flow rate is increased dramatically, a condition that is not advisable in case there is any difficulty on the surface. This is particularly inconvenient in gas wells, in wellhead clusters or wellheads near surface installations, e.g. offshore among other situations.

This is why second generation safety valves were developed, the subsurface safety valves that are controlled from the surface. They are set in the production string at about 30 to 50 meters (100 to 150 ft) below ground level onshore or below seabed (mudline) offshore. They are either directly screwed onto the tubing or set in a special landing nipple. They are connected to the surface by a small, high-pressure hydraulic control line run in along the tubing in the annulus. The control line passes the tubing hanger and comes out of the wellhead to be connected to a control panel that actuates the valve.

3.6.1 Subsurface Controlled Subsurface Safety Valves (SSCSV)

These valves that were often called storm chokes are now termed SSCSV (SubSurface Controlled subsurface Safety Valves). They are set and retrieved by wireline. They close the well following a modification in flow conditions where they are located:

- either when the ambient flow rate increases (and so the pressure loss across the valve also increases), or
- when there is a pressure drop opposite the valve.

The major manufacturers of this type of valve are Otis, Baker and Camco.

3.6.1.1 Pressure differential safety valves (Fig. 3.19)

These valves are designated as "pressure differential valves" or "velocity safety valves" in catalogs and are normally open. A choke incorporated in the valve causes a pressure loss when flowing and this tends to close the valve. A return spring tends to keep the valve open. If the flow rate increases excessively, the supplementary pressure loss that is created induces a closing force that is greater than that of the return spring and the valve closes.

The spring's compression is adjusted in order to set the flow rate (and therefore the pressure drop or the flow velocity) above which the valve closes. The closing and sealing mechanism is either a ball valve, a check valve or a poppet valve.

Because of the way it works, the safety valve closes only if the pressure differential is sufficient. It requires the well it is installed in to produce at a low flow rate compared to the maximum possible output. As a result, its use is confined to certain wells with a high natural

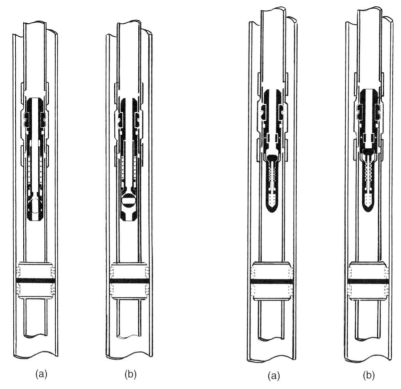

Fig. 3.19 Pressure differential safety valve (*Source:* Otis catalog, 1985).
a. Open. b. Closed.

Fig. 3.20 Pressure operated safety valve (*Source:* Otis catalog, 1985).
a. Open. b. Closed.

potential or to those whose "reduced" flow rate is economically justified, consequently it is not suited to most oil wells. Note also that this safety valve can not tolerate excessively marked flow rate discontinuity (heading), or overly rapid start up after production has been stopped on a well.

3.6.1.2 Pressure operated safety valves (Fig. 3.20)

These valves are also known by the term "ambient safety valves". The closing mechanism is controlled by a return spring and a gas chamber, and the pressure in the well acts to keep the valve open. It is designed to be normally closed.

In order for the valve to open, a pressure equal to or greater than the specified operating pressure is exerted from above. As long as the pressure around the valve remains greater than the set point, it stays open and the return spring is compressed. However, if the ambient pressure drops abnormally, it is no longer high enough to offset the effects of the force exerted by the return spring and the gas chamber. The valve then closes. The closing and sealing mechanism is similar to what was indicated for the preceding type of valve.

Pressure operated valves are well suited to wells whose flow rate is not very sensitive to a variation in the flowing pressure. Some models feature a device that keeps the valves from closing when there are rapid variations in ambient pressure due to the flow instability that is often encountered in certain wells.

3.6.2 Surface Controlled Subsurface Safety Valves (SCSSV)

The SCSSVs (Surface Controlled Subsurface Safety Valves), of the fail safe type, are controlled from the surface by hydraulic pressure in the control line and are normally closed (i.e. closed when no pressure is applied in the control line). The control pressure acts on a jack which pushes a sleeve back thereby opening the valve. At the same time it compresses a powerful return spring. As long as the control pressure is kept at the set operating value, the valve remains open. However, if it falls below a certain threshold then the valve will close automatically solely under the effect of the return spring.

In contrast with SSCSVs, the SCSSV does not depend directly on ambient flow conditions in the well, but rather on one or more parameters measured at the wellhead. This also allows the subsurface valve to be controlled via a number of safety systems connected to process facilities. The well's safety can therefore be achieved manually or automatically whether the trouble is directly related to the well or not: fire, explosion or impact, process problem, etc. Depending on the degree of safety required, the wellhead safety valve alone can be closed or, in conjunction with it, the subsurface safety valve can also be closed.

Due to the design of the hydraulically controlled valve, the depth at which it can be installed in the well is limited by the capacity of its return spring. The spring must be sufficiently compressed to overcome the opposing force due to the weight of the hydrostatic column of the hydraulic fluid in the control line that governs the valve and acts directly on the jack.

Also due to its design, the valve can not be opened as long as the pressure difference between above and below the closing and sealing mechanism exceeds an average of 0.7 MPa (100 psi), so that the valve will not be damaged when it opens. Some valves are equipped with an internal equalization device. Here, all that is required to reopen the valves is to increase the pressure in the control line with the wellhead closed. Otherwise it is also necessary to recompress the tubing above the valve (by means of a pump, a nearby well, etc.).

The SCSSV closing and sealing mechanism is either a flapper valve or a ball valve. For a long time preference was given to ball valves, mainly because of the sealing quality, even though the sleeve weldment (pivot cage) is complex and fragile. Currently check valves seem to be the preferred choice as they are much simpler, more rugged and robust (i.e. safer). Also, a lot of progress has been made regarding the reliability and tightness of flappers.

Conventional SCSSV manufacturers are the same as for the SSCSVs. The valves are offered in two different versions:

- either set in a special landing nipple and retrieved by wireline, in which case they are called "WireLine Retrievable (WLR)", or
- screwed onto the tubing and pulled out with all or part of the equipment in the well, called "Tubing Retrievable (TR)", or "Tubing Mounted (TM)".

There are also special combination valves sold by other manufacturers such as AVA, where one part is tubing retrievable and the other is wireline retrievable.

The SCSSV usually controls only the tubing, but in wells with the annulus full of gas, and especially in wells activated by gas lift, double subsurface safety valves can be installed if necessary. They close both the tubing and the tubing-casing annulus. These SubSurface Tubing-Annulus (SSTA) safety valves also allow the long and dangerous uncontrolled decompression of an annulus full of gas to be avoided (at a wellhead pressure of approximately 7 MPa or 1000 psi and even more).

3.6.2.1 Wireline Retrievable safety valves (WLR) (Fig. 3.21)

The valve is attached to a mandrel that has been modified in order to transmit the control fluid pressure to the valve's jack. Setting the valve and its mandrel in the landing nipple is a tricky job, and positioning and locking it in the landing nipple are crucial. If this is not done properly it may be ejected and rise up to the wellhead the first time it is closed. For enhanced safety in properly setting the valve in its landing nipple, special tools are available on the market based on the following principle:

- a control shear pin is incorporated in the running tool
- on the mandrel, the locking dogs are physically blocked at the end of the setting operation by a sliding safety sleeve that keeps them from retracting
- proper positioning of the safety sleeve is indicated when the control pin is sheared on the running tool.

Installing a WLR valve restricts the inside through diameter of the tubing. In addition, the space allotted to the operating mechanism has been reduced to a minimum so that the inside through diameter is not excessively reduced. As a consequence, the WLR valve is rather complex and fragile. However, most oil wells are equipped with WLR valves because it makes well maintenance easier.

3.6.2.2 Tubing Retrievable safety valves (TR) (Fig. 3.22)

It is advisable to avoid restricting the through diameter in wells with substantial production rates. Additionally, for gas wells especially, this restriction near the surface coincides with a zone where pressure and temperature conditions may be favorable to hydrate formation. This type of well is therefore preferably equipped with tubing retrievable safety valves that pro-

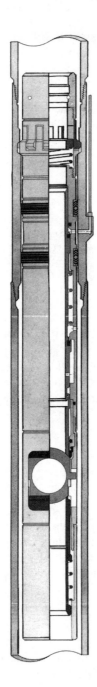
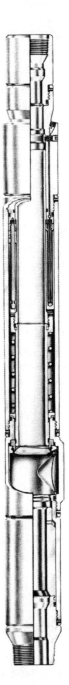

Fig. 3.21 Wireline retrievable safety valve (*Source:* Baker).

Fig. 3.22 Tubing retrievable safety valve (*Source:* Baker catalog, 1984–1985).

vide an inside through diameter the same as that of the tubing. The production string must, however, be pulled out in order to change the valve. The valve is usually associated with what is termed a tubing anchor and a disconnection system that allow only the upper part of the tubing and the valve to be pulled out.

Because of their design, the valves can be locked open:
- sometimes temporarily, an asset during wireline operations in the well, or
- definitely, if the valve should lose its tightness for example.

In the second case, a landing nipple machined in the valve body may allow a wireline retrievable valve to be inserted and controlled by the same hydraulic line. This makes it possible to keep the well in safe conditions while waiting for a workover operation in the event a tubing retrievable valve has failed.

In subsea completions, two tandem valves can be run in together, with each one operated independently of the other.

3.6.2.3 Combination safety valves

Valves that are screwed onto the tubing have an inside cross-sectional area that is compatible with that of the tubing, but require a workover operation when they need to be replaced. Meanwhile, wireline retrievable valves are readily pulled out, but cause a decrease in inside through diameter. A surface controlled subsurface safety valve can be found on the market today that offers approximately the same advantages as both the tubing valve and the wireline retrievable valve (Fig. 3.23). A sleeve, a flapper and a return spring are installed, integrated in the tubing. The jack is off-center and set in the equivalent of a side pocket mandrel. As a result, it is readily retrievable for maintenance by the same technique used for gas-lift valves.

3.6.2.4 Subsurface tubing-annulus safety valves

This system is used in gas lifted production wells, mainly offshore and consists of two valves, one for the tubing and the other for the annulus. Both tubing and annulus can be opened or closed. It requires setting a second packer in the upper part of the well at the same depth as the safety valve (Fig. 3.24). A sliding sleeve allows gas to flow or prevents it from flowing into the annulus by opening or closing the bypass of the upper packer. The sleeve is moved by a hydraulic jack fixed on the tubing and controlled by the same hydraulic line as the tubing safety system.

3.6.2.5 Other safety valves

Subsurface safety valves electromagnetically controlled from the surface, now at the research and development stage, do not require installing any hydraulic control line. If this type of valve proved to be reliable, completion would be simplified and the valve could be set deeper down in the well.

3. THE EQUIPMENT OF NATURALLY FLOWING WELLS

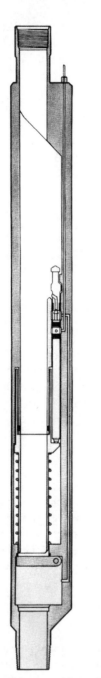

Fig. 3.23 Combination tubing safety valve (*Source:* AVA catalog).

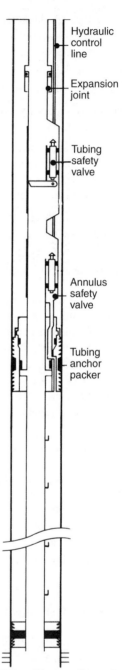

Fig. 3.24 AVA tubing-annulus safety valve (*Source:* After AVA catalog).

3. THE EQUIPMENT OF NATURALLY FLOWING WELLS

3.7 RUNNING PROCEDURE

A large number of factors influence the running procedure, especially the following:
- the type of pay zone-borehole connection, i.e. open hole or perforated cased hole, and for a cased hole, when perforating is done (before or after equipment installation)
- any special operations that might be performed on the pay zone (sand control, stimulation)
- the number of levels to be produced separately in the same well
- the type of equipment chosen, in particular the packer.

We will confine our discussion to proposing a procedure for the following case:

a) Pay zone-borehole connection: perforated cased hole with perforating before equipment installation, no special operations and one single level to be produced.

b) Equipment (Fig. 3.25): includes in particular a permanent packer run in beforehand, a tubing-annulus circulating device and a WLR-type surface controlled subsurface safety valve (SCSSV).

The main differences in procedure due to the use of a retrievable hydraulic packer will be dealt with in Section 3.7.3.

3.7.1 Preliminary operations

Before the equipment is run in several operations are carried out, including the following main ones:
- reconditioning the wellhead
- checking the borehole (tagging cement, cleaning casing walls, displacing completion fluid)
- cased hole logging
- reconditioning blowout preventers (BOPs), if necessary
- re-establishing the pay zone-borehole connection, by perforating.

3.7.1.1 Reconditioning the wellhead

After the production casing (last casing) has been run, cemented and hung, the tubing-head spool is placed on the casing-head spool. It replaces a spacer spool which was incorporated in the drilling wellhead between the casing-head spool and the drilling BOP stack. The BOPs have to be temporarily rigged down so that it can be put into place.

It should be pointed out that in some cases the tubing-head spool is incorporated in the casing-head spool (compact, or solid block wellhead) and so is already in place.

After the BOPs have been reinstalled and the rams have been adapted if necessary to the pipe that will be used later on in the well, the whole system is pressure tested.

3. THE EQUIPMENT OF NATURALLY FLOWING WELLS

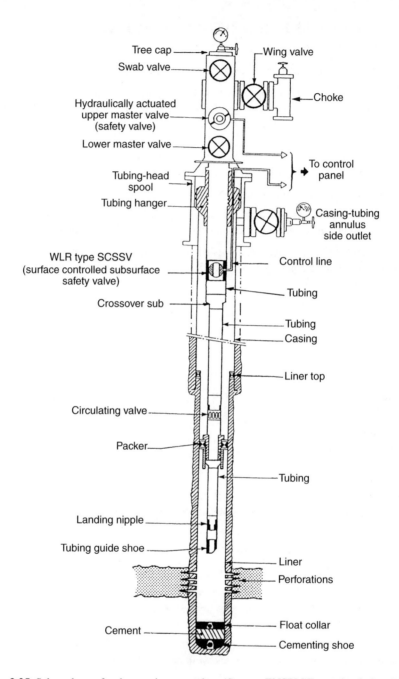

Fig. 3.25 Selected case for the running procedure (*Source:* ENSPM Formation Industrie).

3.7.1.2 Checking the borehole

A. Tagging cement

A drill string is run to check that the well is free of cement down to the float collar after the production casing has been cemented. If it is not, the extra cement is drilled out. In addition, depending on the float collar depth in relation to the pay zone that is to be produced, it may be necessary to drill out the float collar as well and part of the cement below it (see Fig. 3.25).

B. Cleaning the casing walls

Before the completion equipment is installed, a control string is run in the well. It includes a scraper designed to clean the inside of the casing, especially at the packer setting depth where several short trips are made over the same stretch of casing while circulating.

In some instances, particularly when there should be no cement to drill out, the scraper is run in directly on the drill string that is used to check the top of cement.

C. Displacing the completion fluid

With the scraper string in place in the bottom of the well, the equivalent of one or two times the well volume is circulated to recondition the fluid and get rid of any impurities (deposits removed during scraping, etc.). The drilling fluid is replaced by completion fluid, which is usually a clean, solid-free brine (to reduce formation damage during perforating and later on), with a specific gravity sufficient to withstand the reservoir pressure. Supplementary well cleaning can be done by circulating fresh water or sea water before the completion fluid is pumped in.

3.7.1.3 Cased hole logging

Before perforating, it is recommended to check the quality of the cement job. More precisely, for the casing or the production liner, the depth of the cement fill in the annulus should be checked. Any faulty seal in the cement sheath must be detected (due to a poor bond between the cement and the formation and/or the casing or to a lack of cement, among other causes). Depending on the results of the cement bond logs, a decision may be made to attempt a remedial cementing operation.

In order to get a proper reference point in relation to the formation (for the depth of perforations, of the packer, etc.), correlation logs are also recorded and compared with the open hole logs run previously. For correlation, a casing collar locator log, CCL, is usually run along with a gamma ray log that records the natural radioactivity of the formations.

Depending on reservoir engineering needs, other cased hole logs may be run to supplement the ones that were run in the open hole.

3.7.1.4 Reconditioning the BOPs

If there are no existing rams in the BOP stack corresponding to the tubing diameter that will be run, the appropriate rams should be installed and then the BOPs are retested.

3.7.1.5 Re-establishing the pay zone-borehole connection: perforating

We presume that perforating is done by casing gun-type equipment run in on wireline.

The safety of the well is provided first of all by the completion fluid (whose specific gravity is chosen so that bottomhole hydrostatic pressure is greater than the formation pressure) and second by the BOPs.

If multiple-zone completion is planned, the perforations of the upper layers should be scraped to prevent damage to the packing elements of the packers that may be set below them.

3.7.2 Running subsurface equipment in a well equipped with a permanent packer (packer set prior to running the tubing)

The main phases are as listed in order below:
- setting the packer (and its below-packer extension)
- assembling and running the equipment (and testing while running in hole)
- inserting the subsurface safety valve landing nipple
- spacing out the tubing and finishing to equip it
- landing the tubing hanger in the tubing head
- testing the production string (tubing) and the annulus.

3.7.2.1 Setting the packer (and the below-packer extension)

The scraping and circulation during preliminary operations (to some extent) got rid of the particles in the fluid (rock debris, cement, bits of metal, etc.) that might be detrimental to proper setting of the packer (getting it stuck when running, damaging the packing elements, causing inadequate gripping on casing walls, etc.).

When the packer is set on the wireline, a junk basket with a gage ring is always run in first, also by wireline. This is to be sure that the packer will be capable of getting down to the chosen setting depth solely by means of its weight.

The setting depth is chosen according to the cement logs (a zone is chosen where the casing is properly cemented, giving good setting since the casing is held by the cement sheath). Moreover, it is selected so that the tubing shoe is located if possible at least ten to fifteen meters (30 to 50 feet) above the top of the reservoir to be produced (in order to retain the option of perforating further or running production logs over the whole reservoir thickness).

The packer (with an extension screwed on below it) is generally set on the wireline with a setting kit. Since a CCL or a gamma ray are run in at the same time, this allows:

- precise control of the packer setting depth in relation to the reservoir and the cement job
- setting the packer away from a casing collar
- time to be saved as no other round trip with drillpipes is required.

At the end of the setting operation, the setting assembly is automatically released after a release stud is sheared and is then pulled out.

The packer can also be set on drillpipes with an appropriate setting tool. Though the operation is longer and there is less precision in positioning the packer, it is easier to run the packer with certain configurations (highly deviated well or one with a very irregular profile; a packer that is to be set, and therefore placed, in a small diameter liner; a "long" below-packer extension, which is therefore too heavy for the cable). Note that in this case the setting tool is actuated by pressuring up in the pipe. Once the packer has been set, setting can be tested (by tension or by placing weight on it with drill pipe) as well as tightness (pressure increase in the annulus after closing BOP pipe rams). The drill pipe is then pulled out after the setting tool has been unscrewed by rotating to the right.

3.7.2.2 Assembling and running the equipment (and testing while running in hole)

During this phase as well as the following ones, the well is kept under safe conditions first of all by the completion fluid and then:

- for the annulus, by the BOPs by using rams that close on the tubing
- for the tubing, by a valve (or any other suitable closing system) ready to be screwed onto it (i.e. with an appropriate thread and available on the rig floor).

Before starting to run the equipment, the wear bushing (that protects the part of the tubing head where the tubing hanger will be landed) is removed if this has not already been done, and the position of the tie-down screws and their packing glands is checked.

After the locator (or anchor), a tubing (or a short tubing) and the circulating valve (with an integrated landing nipple) have been assembled, the equipment is run in while picking up the tubings.

While the tubings are being picked up, the drift is checked one last time on the walkway. The tubings are handled by an elevator or with a special lifting head and the tubing pin thread protector remains on. The thread protector is removed and the state of the tubing end is verified one last time, it is greased without excess using a grease with the same coefficient as the one used in the factory to screw on the coupling.

After careful alignment, the pin end of the pipe to be screwed on is placed in the coupling of the preceding pipe, preferably using a stabbing guide, especially when the joints are of the premium type. The joint is then connected and tightened to the optimum makeup torque recommended by the manufacturer. This is done carefully and smoothly in order to prevent any over- or undertightening.

Generally speaking tightening to the appropriate makeup torque will theoretically guarantee a good leakproof connection. The problem of tightening is the focus of a lot of attention nowadays and hydraulic tubing tongs are being used increasingly with the torque recorded versus the number of turns. Some examples are the JAM system (Joint Analyzed Makeup) by Weatherford or the Torque Turn by Baker.

While the tubings are being run in the joints can be tested by different methods such as:
- joint by joint testing with water from the outside
- joint by joint testing with gas from the inside
- by section with the completion fluid.

However, the grease that is applied during the connection operation makes the joint leakproof either alone or in conjunction with a sealing shoulder. As a result, unless there is a major defect a prohibitively long testing time would be required to detect a smaller imperfection. This is especially true if the pipe is tested joint by joint. Therefore, these methods are hardly ever used today and great attention is paid to the following instead:
- keeping the pipe threads in good condition before they are screwed together
- monitoring the torque (and the number of turns, etc.) while connecting the joints with care.

Some specific components (circulating device, gas-lift mandrel with a dummy, etc.) are usually pressure tested in the workshop before they are sent to the well site.

3.7.2.3 Inserting the subsurface safety valve landing nipple and continuing to run in

Since a wireline retrievable SCSSV has been chosen, the corresponding landing nipple is inserted between two tubings so that it will be located about 30 meters (100 ft) below ground level (or below the mudline) in the final phase. A separation sleeve (protective sleeve with packings on either side of the hydraulic control line outlet in the landing nipple) is installed and the assembly is pressure tested (unless this was done previously in the shop).

Then the running in operation is continued usually without the control line (also sometimes called a liner) for the time being. It will be connected when the required tubing changes are made for spacing out (see Section 3.7.2.4). Note that sometimes the SCSSV landing nipple is incorporated only during the spacing out phase (particularly for wells onshore — because of the small number of tubings between the landing nipple and the surface, it may be necessary to modify the tubings screwed under or directly on the landing nipple during this phase). The length of the landing nipple must of course be taken into account in spacing out calculations.

3.7.2.4 Spacing out the production string and completing tubing equipment

Spacing out has to be done since the tubing hanger will be screwed onto the tubing instead of a coupling (and does not hold on to the pipe body in contrast to the casing hangers). As a result, a specified length of tubing must be kept between the two fixed points that are the packer and the tubing head.

A. Locating the packer

The number of necessary tubings are not determined by using the packer depth and the length of the different production string components directly. Instead, the packer and its equipment are located by stabbing in the end of the locator or anchor and:
- either by reading the weight drop on the Martin Decker when the locator or anchor seals enter the polished bore of the packer and mainly when tagging the packer, or
- by circulating at a low flow rate in the well to see when the pump pressure rises (and the return flow rate falls) after the locator or anchor seals enter the polished bore of the packer.

If an anchor is used, a tension test is used to check if it is properly engaged in the packer.

B. Spacing out calculations

Before running the equipment, the tubing's breathing (due to variations in production conditions) in particular has been taken into account in order to determine the following:
- for a locator: its total length, how far it should penetrate into the packer and, if need be, the weight of tubing required on the packer when the equipment is set once the tubing hanger is in place in the tubing head
- for an anchor: the weight of tubing required on the packer (or the tensile load exerted on the packer by the tubing) under the same circumstances.

Once the packer has been located as described above, the locator or anchor is set as has been determined (or, in the spacing out calculations, the tubing movement that would be necessary to achieve this situation is taken into account).

The distance between the position of the coupling that is the closest to the tubing head and the position that the tubing hanger will occupy is deduced:
- after having measured the extent to which the last tubing now comes out of the rig's rotary table, and
- according to the distance between the rotary table and the tubing head, as well as the length of the different tubings used.

In order to eliminate this discrepancy in position, the following points are determined:
- which of the last tubings in place have to be removed
- which new tubings and pup joints have to be added to replace them (note that when the tubings were checked and measured on their arrival on the well site, the ones of very different lengths were carefully set aside for this spacing out phase).

C. Spacing out and completing tubing equipment

After spacing out calculations, the tubing is pulled up to the SCSSV landing nipple (or more if necessary for spacing out). When an anchor is used, surface rotation equivalent to ten turns to the right downhole is required in order to disconnect it.

Once the hydraulic control line has been cleaned by circulating an appropriate volume of hydraulic oil through it, it is connected onto the landing nipple by a threaded connection (swage). The assembly is then tested at operating pressure (usually 35 MPa or 5000 psi).

Then the tubing and the control line are run in at the same time:

- after having partially bled off the control line and monitoring the new pressure to detect any leakage (in case the control line has been torn off)
- fastening the control line on the tubing by means of special collars and placing protectors on the couplings
- making the tubing changes required for spacing out.

Once all of the selected tubings have been made up:

- a safety clamp is installed on the last tubing
- the last tubing's coupling is unscrewed and the tubing hanger is screwed on in its place
- the safety clamp is removed
- the control line is connected to the tubing hanger (after a few "dead" turns around the tubing to give it some play) and the assembly is pressure tested
- the landing pipe (used to run the tubing hanger from the rig floor to the tubing head) is connected to the hanger (the landing pipe is made up onto the hanger with moderate torque so that it can be easily unscrewed afterward, but in order to get a satisfactory seal the thread is covered with Teflon tape).

3.7.2.5 Landing the tubing hanger in the tubing head

Running the production string is completed with the landing pipe until the tubing hanger is landed in the tubing head. After having loosened the packing glands, the tie-down screws are screwed on to lock the tubing hanger in place and then the packing glands are tightened again.

3.7.2.6 Testing the production string and the annulus

Once the production string is complete and whatever the method used to test the tubing while running, the string is then tested as a whole (without exceeding the operating pressure of the weakest equipment). Despite a usual testing time of 15 to 30 minutes, the test gives absolutely no guarantee of production string tightness. This is mainly because of the delay in detecting leaks due to the thread grease. To carry out testing, a plug or a simple standing valve (a check valve withstanding pressure from top to bottom) is run in on wireline to the lowest landing nipple. Depending on the equipment used, it may sometimes be preferable to pressure test before the tubing hanger is landed.

An annulus test is also run to check the tightness of the packer (and of the tubing hanger):

- usually at moderate pressure (7 to 10 MPa or 1000 to 1500 psi) so as not to put too much stress on the cement sheath, or
- sometimes according to the casing internal yield pressure (with an added safety factor).

3.7.3 Main differences in running subsurface equipment when there is a hydraulic packer (run in directly on the tubing)

3.7.3.1 Assembling and running the equipment (and testing while running in hole)

The equipment is assembled little by little and in running order. The following will be found for example:

- a below-packer extension including at least one wireline entry guide, one or more tubings, a landing nipple (no-go type), and one or more tubings
- the packer
- a tubing (or short tubing) and the circulating valve (with an integrated landing nipple)
- tubings.

The assembly is run in smoothly so that the packer setting mechanism is not damaged (in other words so that the shear screws that hold the packer in a retracted position are not damaged).

If inside pressure testing is done while running in, the pressure must not be exerted on the packer (for example, test only the part above the packer by setting a test plug in the circulating valve's integrated landing nipple).

3.7.3.2 Inserting the SCSSV landing nipple and finishing to run the equipment

In contrast with the procedure described above (packer set prior to running the equipment), it is not usually necessary to do very precise spacing out. This is because there are no longer two fixed points (the packer and the tubing hanger depth in the tubing head) before the tubing is run in. All that is often done is to refer only to the cumulative length of the components to incorporate the SCSSV landing nipple and then the tubing hanger into the production string at the right time.

Here, once the SCSSV landing nipple has been inserted, the tubing and the control line continue to be run in at the same time. If, however, more precise spacing out is required (packer to be placed between two neighboring pay zones, or opposite a well cemented zone that is not very thick, etc.), bearings can be taken by:

- tagging the bottom of the well (cementing float collar, reference bridge plug) with the string
- running a correlation log versus a radioactive source that was integrated in the equipment when it was run.

3.7.3.3 Partial testing of the production string

Once the tubing hanger and the landing pipe have been screwed on, an "overall" test is run on the production string, though it is limited to the portion above the hydraulic packer (test plug in the landing nipple incorporated in the circulating valve).

3.7.3.4 Setting the hydraulic packer, landing the tubing hanger and testing

The tubing hanger is positioned properly in relation with the tubing head, if this was not done before, in other words:
- it is directly landed in the tubing head, or
- it is placed above it at an appropriate distance (slack off) depending on the compression required on the packer once the packer is set and the tubing hanger is landed.

The packer is then set. A plug (or a standing valve) is set in the landing nipple under the packer (other devices such as a ball that comes to rest in an ejectable or retractable seat can also be used). Then the tubing is pressured up step by step until the packer setting screws are sheared and the packer is set. The shearing and setting pressure depends on the model and the depth but it is usually about 10 MPa (1500 psi). The pressure is held for a while and then bled off gradually.

Before bleeding off, the pressure can be raised up to the selected tubing test pressure, thereby pressure testing the production string as a whole. After the tubing hanger has been landed in the tubing head, the annulus can be pressured up with the same limitations as defined earlier in order to "check" packer tightness.

3.7.4 Installing the Christmas tree and bringing the well on stream

The main phases are as follows in this order:
- the BOPs are replaced by the Christmas tree
- the production wellhead is tested
- the fluids in the well are changed (annular and clearing fluids are pumped in)
- the surface controlled subsurface safety valve is set and tested
- the well is cleared.

3.7.4.1 Replacing the BOPs by the Christmas tree

During all the preceding phases, the BOPs contribute to well safety in addition to the completion fluid. Before the BOPs are disconnected and removed, it is imperative to have at least two (or even three) "safety barriers" in the well in order to maintain a satisfactory degree of safety against blowouts.

Excluding the BOPs and the valve that is ready to be screwed onto the tubing, in the context of the procedure presented here the safety barriers in place are:
- the completion fluid for the tubing and the annulus
- the packer and the tubing hanger for the annulus.

There are therefore enough safety barriers on the annulus. In contrast, the safety on the tubing needs a boost and one or more of the following can be installed:

- a check valve in a profile (usually threaded) specially machined into the tubing hanger, that is commonly called a back-pressure valve, BPV
- one (or more) "plugs" (tight in both directions or only from bottom to top) in one (or more) landing nipple(s)
- possibly, the subsurface safety valve in its corresponding landing nipple.

The less time it takes to replace the BOPs by the Christmas tree, the better the well safety will be, so it is important to prepare the operation very carefully to have the necessary equipment ready (mobilize personnel, operating procedure, hoisting and handling equipment, preparation of the Christmas tree and of its adapter, etc.).

The procedure is as follows:

a) The BOP stack is unbolted and skidded to one side.

b) If there is a riser between the tubing head and the BOPs, it is disconnected and removed.

c) The adapter is mounted on the upper flange of the tubing head. Note that if the tubing hanger is equipped with an extended neck, the adapter also fits on top of it and the following functions are generally fulfilled in this part of the wellhead:
  - seal on the extended neck
  - access by test ports to the "dead" volume between the tubing hanger and its extended neck, the tubing head, the ring gasket and the adapter, and this means that the whole system can be pressure tested
  - continuity of the SCSSV control line between the annulus and outside the well
  - if the well is equipped with a submerged centrifugal pump (see Chapter 4), an outlet is provided for the pump's electric cable.

d) The Christmas tree is installed.

If handling considerations so allow, the different components of the Christmas tree as well as the adapter are very often pre-assembled and tested in the shop. In this way only one connection has to be made on the site.

During the replacement procedure, special attention should be paid to having clean grooves where the ring gaskets are set in place between flanges. Proper cross tightening is mandatory in order to compress the ring uniformly and thereby get a good seal.

3.7.4.2 Testing the production wellhead

A series of hydraulic tests is run as described below:
- Test tightness between the tubing hanger and its extended neck at wellhead operating pressure. Tightness at the top of the annulus can thus be tested at wellhead operating pressure without having to pressure up the annulus (effect on the cement sheath).
- Test the SCSSV control line and particularly its continuity at the wellhead outlet.

- Overall production wellhead test. In order to do this, the BPV previously placed in the tubing hanger is replaced by a two-way check valve (TWCV), tight in both directions.
- If the adapter plus Christmas tree was not pre-assembled and tested in the shop, a valve by valve pressure test is performed (pressure test from upstream to downstream).

3.7.4.3 Pumping in annular and clearing fluids

After the TWCV in particular has been removed from the wellhead, the circulating valve is opened by wireline. The annular fluid (also called packer fluid) and the clearing fluid are displaced into the annulus and into the tubing respectively:

- by direct or reverse circulation
- with a choke on the returns if necessary, depending on the specific gravity of the pumped fluids and on the circulating direction, in order to keep the bottomhole pressure higher than the reservoir pressure.

The clearing fluid is light enough in relation to the reservoir pressure, so that the corresponding hydrostatic pressure at the bottom of the well is sufficiently lower than the reservoir pressure and so that the well can start flowing alone. Water, diesel oil or degassed crude are commonly used as clearing fluids.

If the reservoir pressure is too low to get rid of this type of liquid, the well can be started up by:

- swabbing with the wireline
- nitrogen "lifting", with the nitrogen injected at the bottom of the well by a coiled tubing unit (see Chapter 5, Section 5.3.1)
- if the well does not flow readily enough to produce the specified flow rate all by itself, by using the artificial lift method provided for the well (see Chapter 4).

Once the fluids are in place, the circulating valve is closed by wireline again, imperatively under lubricator protection (see Chapter 5, Section 5.2.1). Its seal may be tested:

- possibly by pressuring up the annulus (but with the limitations mentioned earlier)
- or preferably, by raising the pressure in the tubing, which requires setting and then withdrawing a test plug in the landing nipple under the packer.

3.7.4.4 Setting and testing the SCSSV

With the well still closed at the wing valve, the separation sleeve is pulled out by wireline under lubricator protection and then the subsurface safety valve is run in. Before lowering it into its landing nipple and setting it, the control line is cleared again by pumping hydraulic oil through it. The valve is then set and carefully locked in its landing nipple, and opened by exerting the required pressure in the hydraulic control line. The running tool can then be pulled out.

Since the well is already filled with clearing fluid and is therefore pressurized, the SCSSV is usually tested in the following way:

- the SCSSV is closed by bleeding off the control line pressure down to the tubing wellhead pressure
- the tubing pressure above the SCSSV is (partially) bled off
- the wellhead pressure is observed as long as necessary
- the pressures are equalized on either side of the safety valve either by pressurizing the control line (if the valve has got an equalizing system) or by pumping through the wellhead into the upper part of the tubing
- after pressure equalization, the safety valve can then be opened by using the control line
- the well can then be opened on the surface to proceed to the clearing phase.

The same procedure applies for a periodic test on a SCSSV on a producing well, except that the first step is to close the well on the surface at the wing valve.

The validity criterion for the SCSSV test may vary according to the context it is in. For high-risk wells (high pressure gas, H_2S, highly sensitive environment, etc.) the test is considered satisfactory if there is no increase in wellhead pressure during the observation period (usually about 30 minutes). When there is less risk, one can refer to API recommendations (API RP 14 B) which tolerate a leakage rate as follows for a well open to the atmosphere:

- 400 cubic centimeters per minute for a liquid (i.e. 24 liters per hour)
- 15 standard cubic feet per minute for a gas (i.e. about 25 standard cubic meters per hour).

Other practical rules of the thumb can also be used such as: the test will be considered satisfactory if during the observation period (30 minutes) the variation in wellhead pressure is lower than 10% of the pressure differential exerted on the valve (with a pressure differential of 2 to 4 MPa or 300 to 600 psi).

3.7.4.5 Clearing the well

Since the clearing fluid is already in place, it is enough to simply open the Christmas tree valves for the well to begin to flow. The well is generally produced via a temporary installation including in particular a choke manifold and a flowline connected to a flare and/or a burn pit.

The choke is adjusted so that the flow rate is high enough to clean the bottom of the well and the formation in the vicinity of the wellbore (during the drilling and completion phases, some of the drilling and completion fluids invaded the pay zone to a greater or lesser extent due to the hydrostatic overpressure on the bottom of the hole as compared to the reservoir pressure; if acidizing was performed, the pay zone may still contain spent acid or residues). However, an excessive flow rate should be avoided as it could destabilize the zone in the vicinity of the wellbore or damage the connection between the borehole and the pay zone (especially in the case of sand control).

During the clearing phase, all of the places where liquid is trapped (especially in annuli, the dead volume between the tubing hanger and its extended neck, the SCSSV control line) must absolutely be opened (or their pressure must be monitored) to take the thermal effect into account.

Clearing is performed with the subsurface safety valve in place. However, when there is a risk of damaging the valve by entrained solids, spent acid, etc. the preferred solution is a temporary valve that will be replaced after clearing.

Once the well has started flowing and the effluent is clean enough, it is connected to the normal surface production facilities.

3.7.5 Final completion report

Well completion is not really concluded until the final report has been written. Among other points, the report must include:

- identification of the well
- purpose of the well
- any important facts or events (including trouble) and the results obtained
- the final state of the well
- a detailed account of operations.

For the final state of the well, a detailed technical cross-section must be supplied with the following points in particular:

- the equipment set in the well, its depth and characteristics (nominal diameter, drift, outside diameter where relevant, nominal weight, type of thread and if need be the shear value for the rings or pins to be used when the equipment is removed from the well, and other specific information)
- the characteristics of the annular fluid
- any anomalies such as the presence of a wireline fish (a fish is any tool, piece of equipment, or junk lost or stuck in the well).

The detailed account of operations elaborates on:

- the scheduling of operations and the value of the main parameters (packer setting depth, test pressure, etc.)
- any anomalies observed and any changes made in relation to the initial completion program.

This body of information is extremely valuable, first of all to adapt the completion program of following wells accordingly, and secondly if there should be any trouble or any work done on the well later on. An incident that might seem harmless when it happens can provide the key to understanding and solving a problem that crops up several years later.

Chapter 4

ARTIFICIAL LIFT

Artificial lift allows wells to be produced that are non-flowing or insufficiently flowing. It is mainly designed for oil producers but the technology can also be applied to wells that produce water for a number of different uses such as for utilities or water injection to maintain pressure. It may prove necessary from the beginning of production for oil wells when the reservoir does not have enough energy to lift the fluid to the surface process facilities or when the productivity index is deemed inadequate.

Artificial lift is used in 75% of the wells in the world excluding the US and 90% including the US. There are two main processes:

- mechanical lifting by pumping
- lessening the fluid density by mixing with gas injected in the lower part of the production string, or gas lift.

Note: Do not confuse gas lift, the artificial lift process, with secondary production which consists in injecting gas directly into the reservoir in order to maintain pressure or improve sweep efficiency.

4.1 PUMPING

4.1.1 Principle and types of pumping

In the production string (tubing) that is usually set without a production packer, a pump placed below the dynamic fluid level in the well lifts the crude up to the surface. This energy input allows the fluid to continue on its way and relieves the pay zone of all or part of the back pressure downstream from the pump.

Several pumping techniques can be implemented to help solve the numerous problems that crop up such as: well productivity, type of fluid, completion, on- or offshore siting and environment, as well as economic criteria that must be met.

The two most common pumping methods in the world are:

- **sucker rod pumping:** a downhole positive-displacement pump is actuated from the surface via rods and a reciprocating system
- **centrifugal pumping** by a submerged electric centrifugal pump.

In addition to these two types there is also:

- **hydraulic pumping** by a downhole plunger pump coupled to a hydraulic motor, by jet pump or by centrifugal turbine pumping.

Lastly, but much less common and designed for very viscous crudes loaded with sand and having a high GOR:

- **Moyno rotary pumping** with progressing cavities.

The last two pumping methods are mainly used in the United States.

- As an incidental point we include the vibratory pump designed by Johnston which was used in a lot of big sand-producing wells between 1962 and 1973 in the Chateaurenard region of France on EAP fields. The pump consisted of an oscillator on the surface which was mounted on the tubing head and generated a continuous longitudinal tubing vibration. Check valves placed in each tubing joint allowed the oil which was heavily laden with sand to move up to the surface. A desander incorporated between the wellhead and the flowlines completed the system.

  The system's pumping performance was efficient, but the tubing was quickly subject to fatigue and the oscillator on the casing spool tended to free the casing from the cement sheath with detrimental consequences. Once gravel packing techniques had been mastered on Chateaurenard, this type of pumping was discontinued.

4.1.2 Sucker rod pumping

4.1.2.1 Principle

The system uses a vertical positive-displacement pump consisting of a cylinder and a hollow plunger with a valve. It is run into the tubing screwed onto the end of a rod string. The system is actuated from the surface by a motor that drives a walking beam or a hydraulic elevator (Fig. 4.1).

The pumping cycle can be broken down as follows (Fig. 4.2):

During the plunger's upstroke the plunger valve or travelling valve is closed. The column of liquid corresponding to the stroke will be lifted up to the surface while, relieved of the weight of the fluid, the pressure of the pay zone can then open the bottom valve or standing valve, thereby allowing the pump barrel to fill up with effluent.

During the downstroke the valve of the hollow plunger opens and the standing valve closes, thereby preventing the fluid from returning into the pay zone and allowing the plunger to return freely to its initial point at the base of the pump barrel.

4. ARTIFICIAL LIFT

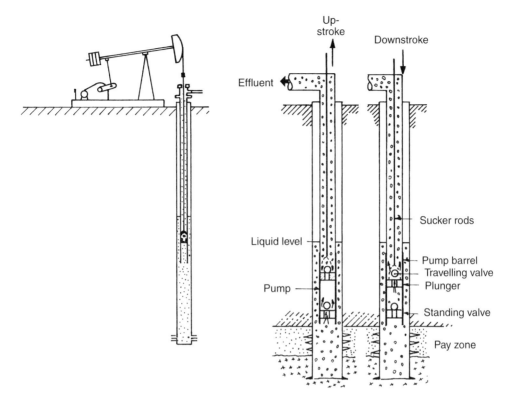

Fig. 4.1 Basic configuration of a rod pumping system (*Source:* ENSPM Formation Industrie).

Fig. 4.2 Rod pumping cycle (*Source:* ENSPM Formation Industrie).

The pump is single-acting and its theoretical output is equal to the volume generated by the plunger's stroke and cross-section multiplied by the pumping rate, i.e. in a homogeneous system:

$$Q = S \times N \times A$$

where:
- Q is the flow rate
- S is the stroke
- N is the number of strokes per time unit
- A is the area of the plunger.

In practice the following parameters are also involved:
- an efficiency factor
- a coefficient depending on the units that are used.

This type of pumping is the oldest of all artificial lift methods, it is simple in design and still widely used the world over.

4.1.2.2 Downhole equipment

This section deals with the following points:

- special equipment for the tubing
- pumps
- pumping rods
- the pumping head.

A. Special equipment for the tubing: anchors and gas anchors

a. Anchoring the tubing

The pumping cycles cause static and dynamic loads on the tubing, and consequently make it move with two effects (Fig. 4.3):

    a) Breathing, a cyclic variation in length due to the weight of the fluid being transferred alternately from the tubing to the rods and vice versa.

    b) Buckling during the upstroke in the part above the pump that is under low tensile load.

This type of stress causes mechanical fatigue in the tubing along with wear and tear on the rod string and couplings. Accordingly, the tubing must meet severe mechanical requirements and it is recommended to anchor the tubing at the pump depth except for shallow, low-flow rate wells. The result of immobilizing the tubing is also to increase the pump's effective stroke at the bottom of the well.

Example of tubing anchoring: using a TM mechanical anchor by Guiberson (Fig 4.4). At the required anchoring depth, rotating the tubing to the left causes the slips to be mechanically pushed out to grip the casing. To unseat the anchor, a little weight is set on the tubing while rotating to the right and the slips retract. If the normal procedure is not enough, the anchor can be released by pulling until specially designed pins are sheared.

b. Gas anchors

When gas is present in the pump it causes inadequate filling and a loss in output. In addition gas lock may occur: when gas remains in the upper part of the pump barrel, the plunger starts its downstroke in the gas with the valve closed and then comes up against the liquid with a sudden impact which is propagated to the rod string. The higher the pumping rate, the greater the impact.

The presence of free gas is mainly related to the bubble point pressure. The problem can be solved by placing the pump far enough below the dynamic level where the height of liquid in the annulus creates a hydrostatic back pressure that is greater than the bubble point pressure. Consequently, this keeps gas from putting in an appearance during the plunger's upstroke.

4. ARTIFICIAL LIFT

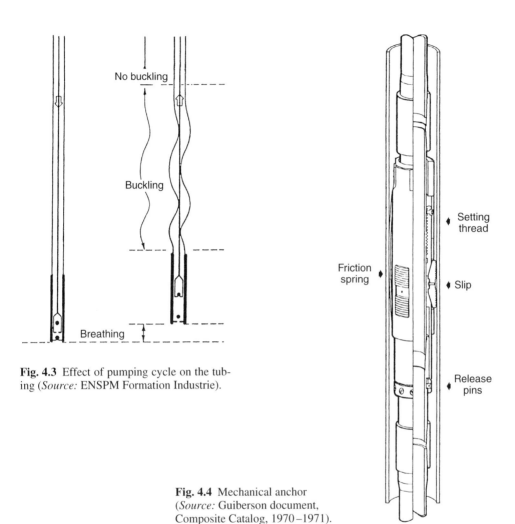

Fig. 4.3 Effect of pumping cycle on the tubing (*Source:* ENSPM Formation Industrie).

Fig. 4.4 Mechanical anchor (*Source:* Guiberson document, Composite Catalog, 1970–1971).

As for the consequences of gas lock, the impacts can be attenuated by reducing the pumping rate, but the stroke or the pump diameter will need to be increased in order to achieve the right output.

If need be, these measures can be greatly improved by a mechanical gas anchor system chosen according to the two types of operation as discussed below:

1) The pump suction end communicates with a pipe that plunges into the base of the tubing that is closed at the bottom and open toward the annulus in the upper part (Fig. 4.5a). The system allows the effluent to separate out, since the oil has to go back down to enter the pump, while the gas continues to migrate freely upward in the annulus.

2) The base of the pump is connected to a special packer which orients the effluent toward the annulus above the packer. The gas can keep moving upward whereas the liquid is taken up in the lower part of the annulus toward the pump inlet.

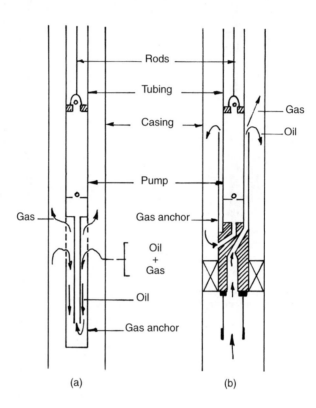

Fig. 4.5 Gas anchor (*Source:* ENSPM Formation Industrie). a. Well without a packer. b. Well with a Page packer.

B. Bottomhole pumps (Figs. 4.6 and 4.7)

Bottomhole pumps are classified by API SPEC. 11 AX into two categories (API: American Petroleum Institute; SPEC: specification):

- R pumps (rod pumps or inserted pumps) which are run complete on the end of the rod string in the tubing and anchored with a seal at the pumping depth in a specially designed landing nipple or sometimes with a packer anchoring device. They can have a fixed cylinder with top or bottom anchoring, or a moving cylinder and a fixed plunger with bottom anchoring (Fig. 4.6).

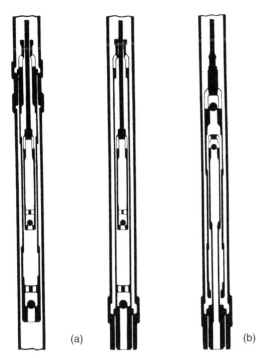

Fig. 4.6 R pumps (*Source:* After an API document).
a. Fixed cylinder and a moving plunger. b. Moving cylinder and a fixed plunger.

Fig. 4.7 T pump (*Source:* After an API document).

Fig. 4.8 Anchoring devices (*Source:* Trico, Composite Catalog 1990–1991).

- T pumps (tubing pumps): the barrel is run first on the tubing. Then the plunger is run in screwed onto the rod string (Fig. 4.7). The plunger has usually got a setting and retrieving mechanism for the standing valve at its base.

The plungers are normally made of metal but can also have synthetic sealing elements. Plunger diameters can commonly range from 3/4" to 4 3/4" OD (outside diameter).

Pump anchoring devices are made of metal or have a cup mechanism, the nipples installed during completion are adapted to the different configurations (Fig. 4.8). Pump valves are systems including a ball and a metal seat inside a closed cage. Plungers can be found with a simple or a double valve (at the top and at the base) for a better seal.

Using the different pumps:

1) R pumps (rod pumps):

- the pump barrel:
  - depending on the depth and the pressure on the pump, the barrel can be one-piece, either thick or thin, or with a liner
- the pump with a fixed cylinder and top anchoring:
  - keeps deposits from settling out in the annulus and avoids unseating difficulties
  - but the main drawback in deep wells is the risk of the pump barrel bursting due to the differential pressure between the inside and the outside of the cylinder
- the pump with a fixed cylinder and bottom anchoring:
  - is recommended for deep wells, as it keeps the pump barrel from stretching during the downstroke, which would be detrimental for a barrel with a liner; moreover the pump is not subjected to bursting stress
  - but the drawback is that deposits may settle out and cause corrosion and problems in unseating
- the pump with a moving cylinder:
  - is recommended when the well produces sand; when the pump stops the valve in the cylinder located in the upper part of the pump keeps the sand from settling out on the plunger
  - however, bottom anchoring is mandatory and this means problems with deposits in the tubing-pump annulus
  - and lastly, this type of pump is not recommended for wells with little submergence because of the considerable pressure losses at the suction end.

2) The T pump (tubing pump) allows a higher output than the R pump but requires the tubing to be pulled out to replace a worn pump barrel.

C. Sucker rods (Fig. 4.9)

These rods are usually cylindrical drawn steel bars with squared off stamped ends for handling and pin shanks fitted with a box coupling on one end.

Sucker rods can also be found now with the rod itself made of fiberglass and glued shanks made of steel.

Fig. 4.9 End of a sucker rod (*Source:* Trico, Composite Catalog 1990–1991).

a. Rod metallurgy

Since sucker rods are subjected to very severe operating conditions (corrosion, vibrations, alternating cyclic stress), they are made of special steels that are alloyed to a greater or lesser extent depending on the required mechanical performance and the environment in the well. API recommends three basic grades of steel which have the following features (Table 4.1):

Table 4.1.

API grade of steel	Chemical composition	Tensile strength (MPa) *(psi)* min	Tensile strength (MPa) *(psi)* max	Recommended use
K	AISI 46XX	586 *85 000*	793 *115 000*	Medium severity loading Corrosive environment
C	AISI 1536	620 *90 000*	793 *115 000*	Heavy loading Fairly corrosive environment
D	Carbon or alloy	793 *115 000*	965 *140 000*	Very heavy loading Non corrosive or low corrosion environment

In addition to these three conventional products, manufacturers can offer special steels that are mainly designed to meet corrosion requirements.

b. Rod sizes

Sucker rods are sized by diameter and are available:
- in 5/8", 3/4", 7/8", 1" and 1 1/8" for normal rods
- in 1 1/8", 1 1/4" and 1 1/2" for polished rods.

Common lengths are 25 and 30 feet, i.e. 7.60 and 9.10 m, but there is also a variety of (short) pony rods that are used for precise string spacing out.

c. Use

1) Making up the rod string

In the well, the rods screwed together make up a rod string. It can be single size or composed of increasingly smaller diameters the deeper down the string goes. The effect of this make up is to distribute the tensile stresses TL/A (*TL*: tensile load on the rods; *A*: rod cross-sectional area) better over the total length of the string between the bottom, which has to withstand only the fluid load (F_o), and the top, which has to withstand the sum of the fluid load along with the weight of the rods in the fluid ($F_o + W_{rf}$) (Fig. 4.10)

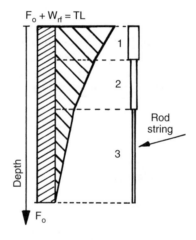

Fig. 4.10 Tensile stresses on a tapered rod string (*Source:* ENSPM Formation Industrie).

Manufacturers supply charts or tables giving the distribution by diameter (four different diameters at the most in a tapered string) expressed in percentage of string length. These recommendations are mainly based on the diameter of the pump plunger, which is in turn optimized according to the tubing diameter, the flow rate, the type of fluid and the pumping depth.

4. ARTIFICIAL LIFT

At the top of the rod string there is a polished rod that is perfectly smooth on the outside and provides the seal for the effluent passing the wellhead by means of a stuffing box. It should be noted that the stroke of the polished rod theoretically represents the stroke of the plunger downhole. In actual fact though, this is not true because, if the tubing is not anchored, its elongations compounded by dynamic inertia phenomena on the rod string (overtravel) will give the plunger and the polished rod a different effective stroke.

2) Resonance in the rod string: choosing the pumping rate

In practice the pumping rate limit is under twenty strokes a minute. In addition, if the pumping rate is synchronized with the natural vibrational frequency of the rods or with a harmonic of the frequency, it can make the plunger bump, rapidly damaging the pump and causing the rod string to fail.

To prevent this type of incident, an asynchronous rate is chosen.

The equation below is used to choose the rate for rod pumping where:

L is the depth of the pump in feet
K is a whole number + 0.5, i.e. 1.5, 2.5, 3.5, etc.
N_{asy} is the asynchronous rate in number of strokes per minute.

a) For a single size rod string:

$$N_{asy} = \frac{237\ 000}{K \times L}$$

b) For a tapered rod string:

$$N_{asy} = \frac{237\ 000}{K \times 0.90\ L}$$

Example: Choose an asynchronous pumping rate for a pump located at a depth of 1500 m driven by a tapered rod string.

- Using the equation:

$$N_{asy} = \frac{237\ 000}{K \times 0.90\ L}$$

with L = 4921 feet (1500 m).
- Results for K = 1.5, 2.5, 3.5, 4.5:

K	1.5	2.5	3.5	4.5
N_{asy}	35.6	21.4	15.2	11.9

- Conclusions: a rate as close as possible to 15 or 12 strokes/minute is chosen.

D. The pumping head (Fig. 4.11)

The pumping head is located on top of the wellhead. It consists of a forged T fixed to the tubing hanger by means of an appropriate extension. The T helps get the effluent out by a side outlet usually consisting of 2" threaded line pipe. On the upper part of the T there is a stuffing box with sealing elements made of graphite or a special compound, held tight with a threaded cap. The sealing elements are tapered and segmented so that they can be placed around the polished rod. The stuffing box can either be screwed onto the T or come together with it as an integral part.

Fig. 4.11 Pumping head (*Source:* Johnson Fagg document, Composite Catalog 1970–1971).

4.1.2.3 Choosing pumping parameters

A. Depth of the pump (Fig. 4.12)

The pump must be located below the dynamic level which depends on:

- the bottomhole flowing pressure (P_{BH}) with:

$$P_{BH} = P_R - \frac{Q}{PI}$$

where:

P_R is the reservoir pressure
Q is the required flow rate
PI is the productivity index
$Q/PI = \Delta P_R$ is the pressure loss in the reservoir and in the vicinity of the well

- the average specific gravity of the produced effluent.

4. ARTIFICIAL LIFT

Note that the reservoir pressure (P_R):

- declines with cumulative production as time goes by
- is also used to determine the static level.

Furthermore, to limit or avoid as much as possible gas entering the pump while the effluent passes the bubble point during plunger upstroke, it is recommended to submerge the pump under a height of fluid in the annulus greater than that corresponding to the bubble pressure at the pump inlet. In practice a safety margin is added to account for fluctuations in the dynamic level (flow impeded in the pay zone-borehole connection, variations in the average specific gravity of the effluent, etc.).

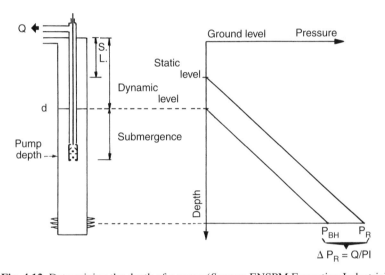

Fig. 4.12 Determining the depth of a pump (*Source:* ENSPM Formation Industrie).

B. Parameters related to flow rate

The pumped flow rate depends basically on the diameter of the plunger, the pump stroke and the pumping rate (or speed). The same flow rate can be obtained by different combinations of these parameters. In the range offered by manufacturers, the choice is made with due consideration given to rod mechanical fatigue in particular.

Fatigue depends mainly on the following:

- number of cycles (pumping rate)
- the difference between the maximum tensile load during the upstroke and the minimum tensile load during the downstroke
- the maximum load in relation to the yield strength.

As a result, it is advantageous to prioritize stroke rather than pump diameter or pumping rate for a given flow rate. Note that the choice of the pumping rate must also take the rod string resonance phenomenon into account as mentioned earlier.

4.1.2.4 Pumping Units (PU)

Pumping units are designed to actuate the rod string for a specified pumping rate and stroke. This requires a mechanical walking beam system or a hydraulic elevator on the surface.

The energy is supplied by an electric motor or an engine (gas or diesel) that is regulated by an adjustable counterweight.

A. *Walking beam units*

The unit consists of the following components (Fig. 4.13):

- an electric motor or engine (gas or diesel)
- a gear reducer connected to the motor/engine by belts and equipped with brakes on the inlet pulley
- a crank and connecting rod system to transform the crankshaft rotation at the reducer outlet into an alternating movement
- a walking beam, supporting the weight of the sucker rods on one side and on the other the adjustable counterweight

Fig. 4.13 Walking beam unit (*Source:* After a Lufkin document, Composite Catalog 1990–1991).

- a horsehead accommodating the polished rod's support cables, and its arcing head which allows the rod to remain in the axis of the well during its stroke
- a Samson post supporting the whole system.

All of the above is set on a skid and rests on a concrete base.

a. Function and adjustment of counterweights

The counterweights are heavy weights placed on the walking beam opposite the sucker rods with respect to the center bearing. Their function is to regulate the output of the motor/engine. Statically, the effect of the counterweight CE is equal to the weight of the sucker rods in the fluid W_{rf} plus half the fluid load F_o:

$$CE = W_{rf} + 0.5 F_o$$

This means a constant stress on the motor/engine that is always in the same direction even though the load changes direction in the horsehead during the pumping cycle (Fig. 4.14).

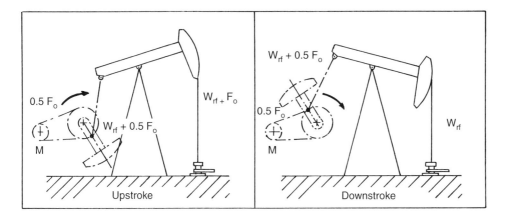

Fig. 4.14 Load on the pumping unit (*Source:* ENSPM Formation Industrie).

In actual fact the load is not static but dynamic and when the counterweights are adjusted due consideration must be given to an impulse factor which depends on the stroke and the pumping rate and will modify the maximum and minimum load on the polished rod. The counterweights are adjusted based on calculations. Furthermore, it is advisable to refine the adjustment on the site using a polished rod load diagram established by means of dynamometric measurement during a complete pumping cycle (see Section 4.1.6: Measurements in a pumped well).

4. ARTIFICIAL LIFT

b. API classification of walking beam units (Fig. 4.15)

1) Standard units

Most manufacturers supply units classified by *API* under the heading B or C depending on the position of the counterweights:

- B unit: the counterweights are on the walking beam (B standing for beam).
- C unit: the counterweights are on the crank (C standing for crank).

The C unit is very widespread and is also termed a conventional unit.

Fig. 4.15 Pumping units (*Source:* ENSPM Formation Industrie).

The B unit is older, simpler and therefore inexpensive, but the C unit has a number of advantages:

- the counterweight adjustment is more sensitive and easier
- the counterweight effect is more modulated since there is less inertia
- the connecting rods always work under tension and the load on the bearings at their head and foot always goes in the same direction.

In contrast, the system is more expensive and requires a larger base.

In addition to the position of the counterweights, the complete classification includes performance criteria, characterized by three numbers:

- the maximum torque at the reducer outlet in thousands of inch pounds
- the maximum load on the polished rod in hundreds of pounds
- the maximum stroke of the polished rod in inches.

Some manufacturers such as Lufkin or Mape who supply two-stage reducers add the letter D after the number that stands for the torque.

4. ARTIFICIAL LIFT

Examples:

Lufkin unit: C 456D - 256 - 120

- counterweight on the crank
- maximum torque at the reducer outlet (double): 456 000 inch pounds
- maximum load on the polished rod: 25 600 pounds
- maximum stroke of the polished rod: 120 inches.

Mape unit: BB 57D - 109 - 48

- counterweight on the beam
- maximum torque at the reducer outlet (double): 57 000 inch pounds
- maximum load on the polished rod: 10 900 pounds
- maximum stroke of the polished rod: 48 inches.

Note that the same type of classification is found for the other types of units used in rod pumping.

2) Special units

In addition to these two conventional types of units, some manufacturers supply units that are better adapted to specific problems. This is the case for Lufkin, which supplies a unit called Unitorque designed for a better distribution of engine torque, and another called Air Balanced with a long stroke designed to increase the flow rate.

Lufkin Mark II Unitorque pumping unit (Fig. 4.16)

The geometry of the crank and connecting rod system has been modified on this unit:

- The beam and its support are articulated behind the reducer.
- The upper fixed point of the connecting rod is very near the horsehead, and this gives an approximate crank rotation angle of 195° in the upstroke and 165° in the downstroke.
- The counterweight is always on the crank but offset by 15° in the rotation direction.

When the unit runs counter-clockwise, the geometry of the system creates an offset of the counterweight top and bottom dead center in relation to the sucker rods. This gives more evenly distributed and lower torque at the reducer outlet for the same stroke length.

Lufkin A or Air balanced unit (Fig. 4.17)

The counterweight effect is provided by accumulated air at a specified pressure in a moving cylindrical chamber (cylinder + air receiver tank) connected to the pump beam and equipped with a piston fixed to the unit's base frame.

Not having a counterweight means that:

- The unit is lighter and can be trailer mounted (for well testing).
- Longer strokes are possible, up to 300" (7.60 m).

However, an air compressor must be installed to offset leakage and maintenance costs are rather high.

4. ARTIFICIAL LIFT

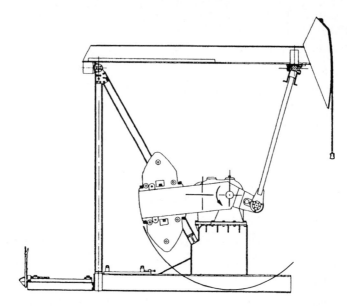

Fig. 4.16 Lufkin Mark II Unitorque pumping unit (*Source:* After a Lufkin document, Composite Catalog 1990–1991).

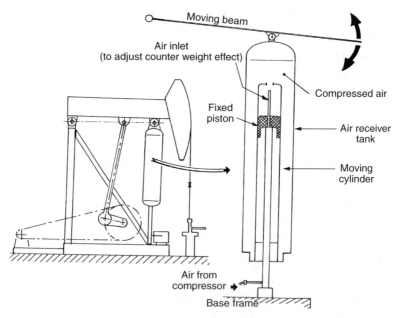

Fig. 4.17 Lufkin A (Air balanced) pumping unit (*Source:* After a Lufkin document, Composite Catalog 1990–1991).

4. ARTIFICIAL LIFT

B. Hydraulic units

In this type of surface unit, the hydraulic energy of pressurized oil is used to drive the sucker rod string:

- either by rolling and unrolling a winch, or
- by hoisting with a piston jack.

These units are more complex in design than beam units, but are nevertheless quite advantageous offshore because of their performance despite limited floor space and bearing capacity.

The French manufacturer Mape has designed and marketed the following two types of units:

a. The Long Stroke H unit (with a hydraulic winch) (Figs. 4.18a and 4.18b)

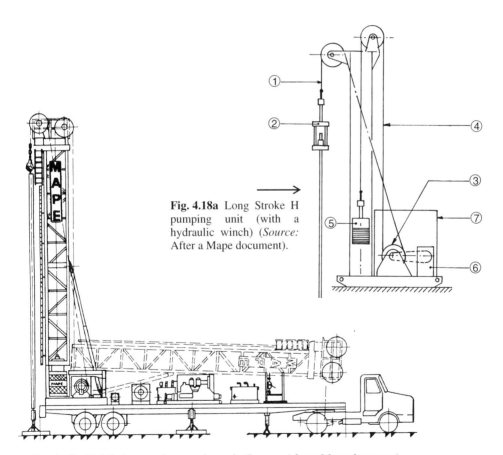

Fig. 4.18a Long Stroke H pumping unit (with a hydraulic winch) (*Source:* After a Mape document).

Fig. 4.18b Mobile long stroke pumping unit (*Source:* After a Mape document).

4. ARTIFICIAL LIFT

The unit is in the shape of a cylindrical column or a square tower equipped with two pulleys. The polished rod is hung on two cables (1) by means of an articulated rod hanger (2). The cables are rolled up around the drum of a winch (3) and opposed to two other cables (4) which support the balancing counterweight (5). This is all run by a motor that always goes in the same direction at a constant speed and drives a hydraulic motor (7) connected to the winch drum via a variable displacement pump. An inversion mechanism (6) driven by the winch automatically triggers the direction in which the winch rotates at the polished rod's top and bottom dead center. The pumping rate can range from 0 to 5 strokes per minute and is adjusted by changing the pump output. The stroke can go up to 394", i.e. 10 m, by regulating the inversion on the winch.

Mape also builds mobile units based on this system that are ideal for well testing.

The ten-meter stroke allows the unit to pull its own rods out, thereby making further workover jobs unnecessary.

b. The cylinder rod pumping unit (Fig. 4.19)

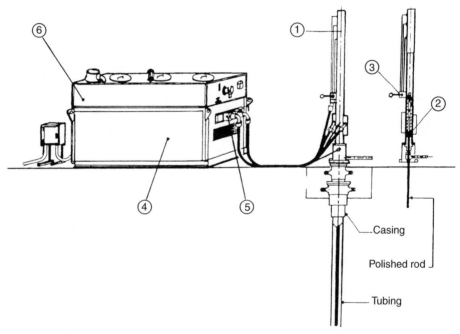

Fig. 4.19 Cylinder rod pumping unit (*Source:* After a Mape document).

This unit uses a hydraulic jack to move the sucker rod string. It consists of a jack (1) and its piston (2) connected to the polished rod and a stroke selector (3) in the axis of the well. The jack is connected by high pressure hoses to a hydraulic unit (4) with:

- an electric motor or engine (electric, diesel, gas, etc.)
- a hydraulic pump with a pumping rate selector
- a pilot slide valve (5) to transfer the oil from one side to the other of the piston
- an oil tank (6), accessible by a weighted valve for the return
- a pressure accumulator that acts as a counterweight: during the upstroke it adds its effect to the hydraulic pump and is inflated again by the pump during the downstroke.

The manufacturer can provide hydraulic units that can activate one or two wells for cluster wells. Recently the company has started selling a new unit of this type, with a much longer stroke (360" instead of 144") based on a different hydraulic scheme. In particular, it uses a variable displacement pump that can reverse the flow gradually and smoothly, thereby lessening the fatigue and impacts on the system. Additionally, the energy efficiency of the installation is better.

4.1.3 Submerged centrifugal pumping

The centrifugal pump is run in to its depth in the casing on the end of the production tubing. Electric energy is supplied to the motor by a cable that is unrolled and clamped to the tubing while running in.

On the surface a special tubing head with packing elements allows the cable to come out to a control panel. An adjustable choke serves to adjust the flow by increasing or decreasing the back pressure on the pump.

If the downhole unit should fail, the tubing and pump should be pulled out together for repairs.

4.1.3.1 Component parts of the centrifugal pump (Fig. 4.20)

Schematically, the pump set in the well consists of three main parts connected together and aligned along the same axis. They are, from bottom to top:
- the electric motor
- the protector or seal section
- the pump with its intake system.

A. *The pump* (Fig. 4.21)

This is a multistage centrifugal pump with the number of stages — sometimes up to several hundred — needed to get the specified pressure head. Each stage consists of a rotary impeller (paddle wheel) that supplies energy in the form of velocity to the fluid to be pumped and a static diffuser that transforms the kinetic energy into pressure energy before sending it to the impeller located just above. The stages are stacked up inside a liner.

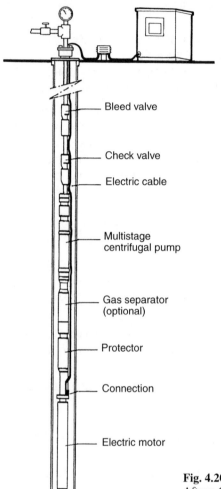

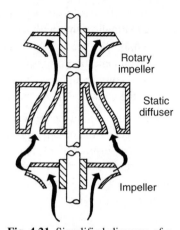

Fig. 4.21 Simplified diagram of a centrifugal pump stages (*Source:* ENSPM Formation Industrie).

Fig. 4.20 Electric pumping equipment configuration (*Source:* After a Centrilift document, Composite Catalog 1990–1991).

The pressure head provided by one stage depends on the diameter of the stage, the geometry of the impeller and the diffuser and the rotary speed of the rotor. Note that in a centrifugal pump the delivery head is independent of the fluid that is pumped.

The fluid coming from the annulus enters the pump through the intake section. This section can be coupled upstream with a static baffle type or centrifugal type downhole gas separator if there is any free gas present. The separator is designed to shunt a large part of the gas toward the annulus to prevent pump cavitation and reduce load fluctuations on the motor. The pump outlet or discharge head has got a coupling with an adapter that is screwed directly onto the tubing.

Above the pump there is a check valve and a bleed valve that can be opened by dropping a sinker bar. These accessories serve:

1) to prevent the pumped fluid from flowing back down during shutdowns,
2) to pull out the pump with the tubing empty.

B. The electric motor

This is a squirrel cage induction motor, two poles and three phases, that usually runs at about 2900 rpm for 50 Hz and 3500 rpm for 60 Hz (US), at a voltage ranging from 600 to 2400 volts and variable intensity from 20 to 120 amperes. It is enclosed in a steel housing full of oil that lubricates the bearings, isolates the motor electrically and cools it by transferring heat outside to the effluent coming up in the well.

Because of this the motor must never be placed below the perforations, i.e. below the zone where the produced fluid flows up the well. Moreover, an inadequate fluid flow rate can cause abnormal heating and damage the motor.

Electric energy is transmitted to the motor by a specially designed cable, round or flat on the outside with conductors inside a steel or Monel sheath. The round cable is normally recommended but the flat one can be used when the space available between the pump and the casing would cause problems during installation.

A range of conductors covers the needs of the different motors. As for the temperature range, the motor and cable are usually designed for use up to 150°C (300°F), but the limit can be pushed up to 400°F (205°C) by a special cable sold by Centrilift.

C. The protector

Also called a seal or an equalizer depending on the manufacturer, the protector provides a tight connection between the motor and the pump. It has a chamber filled with clean oil that creates a pressure equalizing "pad" between the motor oil and the well effluent, while seals keep the oil from migrating along the shaft. The protector also includes a special thrust bearing that handles the pump's axial thrust so that it is not transmitted to the drive shaft.

4.1.3.2 The surface control equipment

In addition to transformers, manufacturers also supply a control panel for each pump that includes apparatuses and devices to protect the motor and provide flexibility of use such as: ammeter-recorders, overload relays, circuit-breakers, stop and re-start timers.

A system has been on the market now for a few years that is used to vary the motor speed and consequently the pump output rate automatically. This flexibility allows wells to be operated better when reservoir pressure, productivity index, GOR (gas/oil ratio) and WOR (water/oil ratio) are not stabilized.

4.1.3.3 Selecting a pump (Fig. 4.22)

In principle, the centrifugal pump has little flexibility and requires the most thorough understanding possible of present and future well performance: in particular the productivity index, the bubble point, the flow rate, and the pressure head which includes the pressure loss by friction in the tubing and the required wellhead pressure. This total pressure head is termed total dynamic head, TDH.

The choice of the pump will be restricted by the casing diameter first of all.

Then the pump and motor are chosen according to the performance indications published by manufacturers and presented in the form of curves giving the flow rate, the pump efficiency and, for one stage, the delivery head and the power absorbed for a fluid with a specific gravity of 1 (Fig. 4.22). The delivery head per stage depends on the diameter of the system and on the geometry of the impeller and of the diffuser. The number of stages required is the TDH divided by the delivery head per stage.

Finally, the rated power of the motor is calculated by multiplying the maximum power per stage taken from the pump curve by the number of stages and correcting for the specific gravity of the pumped liquid.

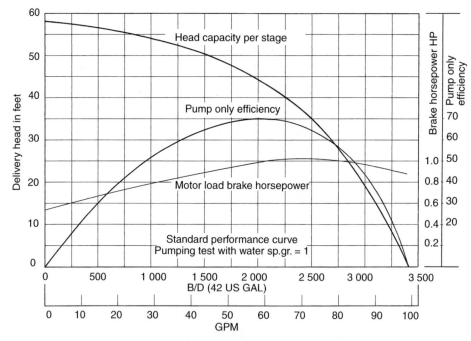

Fig. 4.22 Performance curve for a centrifugal pump (*Source:* document from the "artificial lift" seminar, ENSPM Formation Industrie).

4. ARTIFICIAL LIFT

4.1.4 Hydraulic pumping

Artificial lift by hydraulic pumping was first implemented in 1930 in the United States. It is widely used both there and elsewhere in the world, but has not been very popular in France to date.

4.1.4.1 General introduction

Hydraulic pumping is an application of Pascal's principle, which shows that liquids transmit the pressure variations that they undergo integrally and at all points. Hydraulic pumping applies this principle to artificial lift by transmitting pressure generated on the surface to the bottom of a well by a working fluid in order to actuate:

- either a piston that drives a positive-displacement plunger pump by means of an alternating movement
- a jet pump equipped with a choke that leads into a venturi, in order to carry the fluid from the pay zone together with the working fluid
- or a turbine pump where a turbine drives a centrifugal pump.

The working fluid can either be oil from the reservoir itself, or the reservoir water or any other compatible fluid, in particular a lighter oil or a fluid containing a fluxing product for example.

Historically speaking, the plunger pump came into being in 1930, before the jet pump. It was only after the Second World War that the jet pump was used to lift production from oil wells economically because of the progress made by American manufacturers in the materials and manufacturing of the pump's main parts. The turbine pump is a more recent development and is still not very widely used.

4.1.4.2 Principle of the plunger pump (Fig. 4.23)

Schematically speaking, the hydraulic pump consists of two connected pistons mounted on the same axis, with each one moving in a cylinder.

The upper one is the engine piston and is actuated by the power fluid that comes from the surface via a feed tubing. A engine valve sends the fluid alternatively into the lower or the upper chamber of the engine cylinder.

The pump piston, connected to the engine piston, sucks the well effluent in and discharges it on the surface. The pump is equipped with four check valves and is double acting. Note that an equalizing conduit along the plunger axis with its outlet in a closed chamber (balance tube) located under the lower pump cylinder allows the moving part to work under equalized pressure.

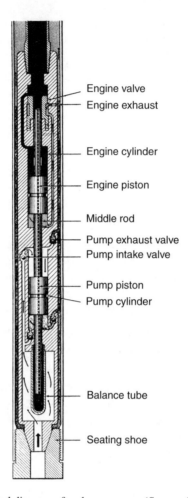

Fig. 4.23 Simplified diagram of a plunger pump (*Source:* Kobe document, Composite Catalog, 1972–1973).

Generally speaking, i.e. in 90% of the cases, the power fluid and the produced oil are pumped out together through the same production piping. This is termed the open power fluid system, OPF. However, the remaining percentage is the closed power fluid system, CPF, where the power fluid and the produced fluid are pumped out separately to the surface. The second system involves more equipment, but makes well testing easier and facilitates production metering since the two fluids are not mixed together.

In practical terms, American manufacturers such as Kobe or Trico Industries offer several different assemblies based on the system described above. They may sometimes have a driving plunger diameter different from the pump plunger, a pump with a double driving plunger or a double pump plunger, or with a large displacement. This gives a wide range of depth and flow rate uses for common tubing diameters, i.e. from 2 3/8" to 4 1/2" OD.

4.1.4.3 Well equipment

A. *Installing the pumps* (Fig. 4.24)

Two types of completion are used to install the pump in the bottom of the well:

- **fixed pump**
- **free pump.**

The fixed pump is run in screwed to the base of the feed tubing, the whole system concentrically inside either the production tubing or the casing. The produced fluid and the working fluid are pumped up to the surface through:

- the production tubing in the **fixed insert** configuration (Fig. 4.24a)
- the casing in the **fixed casing** configuration, where the feed tubing and pump are used in conjunction with a packer (4.24b).

In the free pump system, which is becoming more widespread, the pump is used like a pumped tool. The upper part is equipped with a fishing neck and an elastomer cup that allows the pump to be set and retrieved by circulating.

Two choices are possible: free parallel and free casing.

The **free parallel** option (Fig. 4.24c) has two tubings, one for the working fluid and the pump and the other to evacuate the fluids (produced effluent and working fluid) to the surface. The two tubings are run in side by side and a connection is made between them by a crossover block system equipped with a foot valve at their base.

In the **free casing** option (Fig. 4.24d) the base of the tubing is anchored in a packer and the pump is set in a landing nipple that also acts as a foot valve. The working fluid comes down through the tubing, the effluent and the working fluid go up through the casing-tubing annulus.

These four types of completion of varying degrees of complexity give some leeway to adapt to specific production requirements such as: easy operations on the well, flow rate range and presence of free gas.

B. *Particular wellhead features*

With the fixed pump system, the wellhead is extremely simple. Above the tubing-head spool there is a connecting flange connected to:

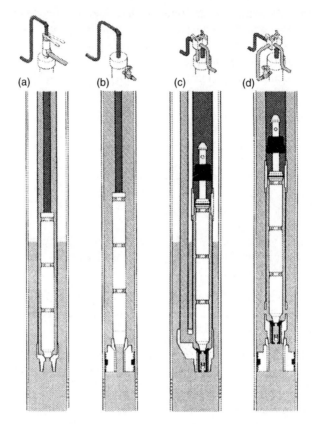

Fig. 4.24 Types of pump classified according to how they are installed in the well (*Source:* Trico document, Composite Catalog, 1990–1991).

- a production outlet T, with the small tubing for the working fluid passing through the assembly in the fixed insert configuration
- the working fluid feed pipe in the fixed casing configuration.

In contrast, the free pump type of completion has a wellhead equipped with a special device based on a four-way valve. It is used for direct or reverse circulation as required to run in or pull out the pump, in addition to pumping the effluent normally. A lubricator is also mounted temporarily on the wellhead to make tool pumping operations easier.

C. Surface facilities (Fig. 4.25)

The hydraulic pumping surface facilities basically work in a closed circuit. The main part is a positive displacement pump that pumps the working fluid into the well after sucking it up from a tank that holds the well's produced fluid. In practice the pump is of the triplex positive displacement type that can give high pressures of approximately 7 to 35 MPa (1000 to 5000 psi) as required in hydraulic pumping.

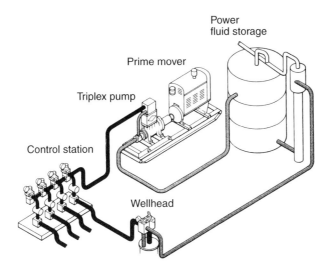

Fig. 4.25 Surface facilities for hydraulic pumping (*Source:* After a Trico document, Composite Catalog 1990–1991).

For each type of pump there are several plunger diameters that cover a wide range of flow rates — the working fluid and produced fluid flow rates are of about the same order of magnitude. The flow rate can be adjusted by recycling part of the fluid via a relief valve placed on the pump. The working fluid comes from a storage tank located in a nearby production installation or from buffer storage located at the well outlet at the beginning of the production flow line.

The pump can function either on one single well or on a distribution manifold connected to several wells. When several wells are involved, the pressure and flow rate of each takeoff downstream from the manifold are adjusted by an adjustable choke. The input circuit also includes supplementary equipment such as pressure safety valves, an oil level, a cyclone separator and a metering pump to inject chemicals, so as to inject the cleanest possible working fluid with the best possible reliability.

D. Hydraulic plunger pump performance

The common range of pumps available on the market give outputs from 20 m^3/d (120 bpd) to 800 m^3/d (4800 bpd) with delivery heads going from 1500 m (5000 ft) to 4500 m (15 000 ft).

Kobe has a special free casing type E pump with a 2 7/8" pump plunger that is run inside a 4 1/2" tubing and can give an output of 1250 m³/d (8000 bpd) with a pressure head of 2600 m (8500 ft).

Tubing OD (inches)	Pump plunger stroke (inches)	Pump rate (strokes/ minute)	Pump plunger diameter		Output (m³/d)	Delivery head (m)
			Single barrel (inches)	Double barrel (inches)		
2 3/8	12	121	1 3/16		22	4 500
3 1/2	24	87	1 3/4		190	3 000
				1 3/4	380	1 500

The table above illustrates performance figures for two Kobe pumps marketed by Trico Industries (specifications taken from the Composite Catalog 1988–1989). The two pumps are of the casing free type, one is set in a 2 3/8" OD tubing and the other in a 3 1/2" OD tubing.

4.1.4.4 Principle of the jet pump (Fig. 4.26)

The working fluid gets to the nozzle under high pressure, with the pressure energy then transformed into kinetic energy at the nozzle outlet where it comes into contact with the produced effluent. The effluent is carried along by the high velocity of the working fluid. The two fluids get to the throat together then to the diffuser, which transforms the kinetic energy of the working fluid-well effluent mixture into pressure energy, thereby allowing both fluids to reach the surface together.

Completion is usually of the free casing type. The pump is run into the well in a tubing and set in a landing nipple fixed to a packer set in the casing, with all components constituting the bottom hole assembly. The pumped effluent comes up through the annulus together with the working fluid.

Completion of the free parallel type is also possible but is less common.

The main advantages of this pump are:
- high flow rates and GOR
- high mechanical reliability.

Its major drawback is its low efficiency (25 to 30% compared to 65 to 70% for plunger pumps).

4. ARTIFICIAL LIFT

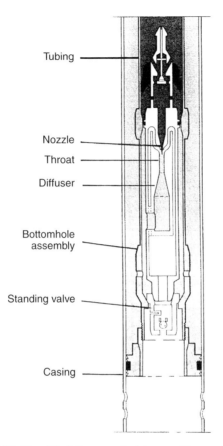

Fig. 4.26 Simplified diagram of a jet pump (*Source:* Trico document, Composite Catalog, 1990–1991).

4.1.4.5 The turbine pump

The principle is that the working fluid injected into the well makes a hydraulic turbine rotate, which in turn drives a centrifugal pump. The diagram (Fig. 4.27) shows one possible configuration with the working fluid injected through the tubing and the working fluid and produced oil returning through the casing.

The manufacturer Weir Pumps Ltd. supplies the following information:

- Pump output can range up to several hundred cubic meters a day (several thousand barrels per day).
- Turbine power can range from 7.5 to 750 kW and the pump can be set in the well at depths in excess of 3000 m (10 000 ft) with no trouble.

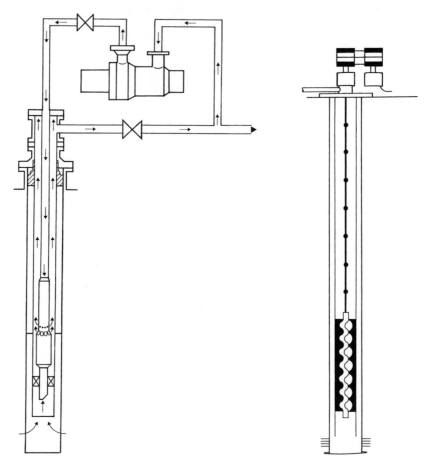

Fig. 4.27 Simplified diagram of an installation with a turbine pump (*Source:* Weir Pump Ltd. document, Composite Catalog, 1986–1987).

Fig. 4.28 Simplified diagram of an installation with a Moyno pump (*Source:* After a Rodemip document).

4.1.5 The Moyno pump (Fig. 4.28)

This type of positive displacement pump with progressing cavities is the result of research by the French engineer René Moineau in 1932.

It works in the following way:

- A rotor consisting of a steel helicoid screw rotates inside a molded elastomer stator that has a double helix inner profile.

- The rotation generates cavities or cells that progress along the axis, which sweeps the fluid upward in a vertical pump with a flow rate that is directly proportional to the rotational speed of the helix.

This system is marketed in the United States under the name Moyno by Robbins and Meyer and has recently been developed under the name Rodemip by PCM Vanves (France). In the traditional version the pump is driven with an adjustable speed from 100 to 500 rpm by a rod string. The axis of the string is fixed to a rotating head on the surface, which is mounted on bearings and connected to an electric motor by a belt.

The advantages of this type of pump are those of a screw pump that can handle just about anything:

- viscous fluids
- fluids with high concentrations of paraffin
- fluids where gas is present.

However, the problems due to the rotation of the rod string make it ill suited to high flow rates, highly deviated wells and excessive depths (over 1200 m or 4000 ft).

4.1.6 Measurements on pumped wells

There are three types of measurements:

- liquid levels in the annulus
- dynamometric for rod pumping
- amperage readings for submerged centrifugal pumping.

The location of the level in the annulus of a well without a packer is used to monitor:

- at a standstill: the height of fluid that is representative of the changes in reservoir pressure according to its average specific gravity
- while running: the dynamic submergence of the pump, measured by means of a sonic depth finder or echometer.

To check that rod pumping units with mechanical counterweights are running properly, dynamometric measurements must also be made at the polished rod.

Lastly, amperage recorders can be found on the control panel of submerged centrifugal pumps. They are representative of the way the electric motor is running and are consequently indicative of pumping conditions.

The following section deals with level and dynamometric measurement techniques which are representative of rod pumping. The use of intensity diagrams in centrifugal pumping is also discussed.

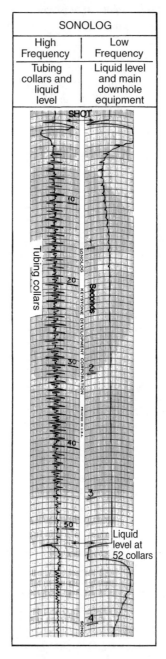

Fig. 4.29 Sonolog recording (*Source:* After a Keystone document. Composite Catalog, 1990–1991).

4. ARTIFICIAL LIFT

4.1.6.1 Measuring levels in the annulus (echometer)

The level of liquid is located by the reflection of a sonic echo on tubing joints and finally on the level of the liquid itself. It is recorded on the surface in the form of electric pulses on a strip of paper that unrolls as a function of time.

The echometer consists of two main parts connected by a cable:
- a microphone
- a receiver-recorder.

A. *The microphone*

Screwed onto an elbow pipe at the entrance to the casing, it converts the sonic echo into electric current. The sonic echo comes from a chamber that opens onto the casing and:

- a cartridge filled with an explosive can be fired into the chamber by means of a hammer
- a sudden discharge of compressed gas (CO_2) can be sent into it by means of an air valve from a reservoir bottle
- or if the casing is under gas pressure (a minimum of 200 psi) the gas can be discharged into a specially designed auxiliary chamber.

B. *The receiver-recorder*

Placed in a case, it amplifies and records the electrical indications on an electrosensitive paper diagram that unrolls as a function of time at a velocity appropriate to the velocity of the sound wave. It works by rechargeable battery or an independent 12 volt battery. Interpretation is made easier by the use of an adjustable range selector that can "read" intervals of 10 joints of tubing.

Note: Keystone, the US manufacturer, supplies an echometer called Sonolog with a two-track receiver-recorder, one of which gives all the reflections from the joints, and the other more attenuated one indicates only the important echoes such as when the shot is fired, the liquid level, the perforations, the tubing anchor, etc. (Fig. 4.29). The two Sonolog amplifiers have different responses that are designed to separate the two frequency characteristics properly. This gives clear recordings at great depths and even under difficult conditions such as the presence of paraffin or a partially empty casing.

4.1.6.2 Dynamometric measurements at the polished rod

A. *Purpose*

It is necessary to make dynamometric measurements at the polished rod, i.e. at the junction point between the surface unit and the pumping string, in order to monitor a beam unit rod pumping installation effectively. A loading diagram is recorded during a pumping cycle to check and correct if need be the parameters that are important for proper operation such as:

- the maximum and minimum weight on the sucker rods compatible with allowable mechanical stress
- the effects of the counterweight on the unit and, by deduction, the torque on the reducer outlet during the upward and downward movement, as well as the absorbed power.

In addition to these "fixed" points, the diagram pattern is also used to interpret how the pump works downhole and detect any anomalies such as:

- valve leakage
- presence of gas in the pump
- broken or unscrewed sucker rods
- unseated R pumps
- partial sticking due to sand.

B. Theoretical loading diagram pattern (Fig. 4.30)

The combination of static stress (weight of fluid and sucker rods) and dynamic stress (alternating traveling movements of variable velocity and acceleration, along with elasticity and inertia effects) on the sucker rods can give a loading versus movement diagram that looks like the one below for a pump filled 100% and a medium pumping rate without any resonance effect.

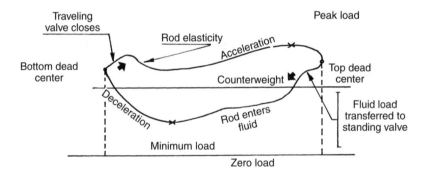

Fig. 4.30 Theoretical dynamometric diagram pattern (*Source:* Artificial lift seminar document, ENSPM Formation Industrie).

Also to be noted in addition to the loading diagram is the counterweight line plotted when the unit is at a standstill. The rod string was immobilized with chains when the counterweight effect was at a maximum (horizontal beam or crank).

Theoretically:

$$CE = \text{counterweight effect} = \frac{\text{peak load} + \text{min. load}}{2}$$

Reminder:

$$CE = \frac{\overset{\text{peak load}}{\overset{\uparrow}{W_{rf} + F_o} } + \overset{\text{min. load}}{\overset{\uparrow}{W_{rf}}}}{2} = \frac{2W_{rf} + F_o}{2} = W_{rf} + 0.5\, F_o$$

In this way the torque at the reducer outlet can be checked and rebalanced if necessary by readjusting the counterweights.

The various production anomalies on the pump and the harmonic resonance phenomena on the sucker rod string can give a very different image from the theoretical diagram, as illustrated Fig. 4.31.

C. The principle of dynamometers

The measurements are made on the polished rod by a dynamometer that has been temporarily placed between the horsehead connecting crossbeam and the polished rod clamp and can be performed by:

- a portable compact Leutert hydraulic type dynamometer with a pen that records the measurement on a diagram (Fig. 4.32), or
- an electronic cell dynamometer on the polished rod with remote recording of the digitized reading on a printer (Delta X type for example).

Dynamometers are highly accurate and can also keep the measured parameters on record for computer processing.

4.1.6.3 Measuring intensity in submerged centrifugal pumping

A. The principle

A amperage recorder is incorporated on the pump control panel on the surface. It indicates the intensity I of the current in the motor versus the elapsed operating time. The diagrams are interpreted in order to monitor the pump's operating conditions.

Note that the hydraulic power supplied by the pump is mainly dependent on the flow rate, the delivery head and the specific gravity of the pumped effluent. The electric power P absorbed by the three-phase motor that drives the pump is in the following form:

$$P = UI\sqrt{3} \cos \alpha$$

where U: voltage at the motor's terminals = constant.

As a result, P/I is constant overall, and the power and the intensity always vary in the same direction. This is the basis for interpreting amperage diagrams. Four typical diagrams are analyzed below.

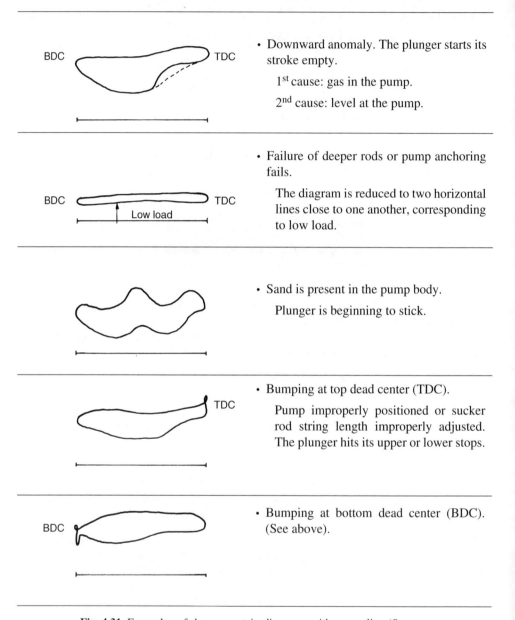

Fig. 4.31 Examples of dynamometric diagrams with anomalies (*Source:* ENSPM Formation Industrie).

Fig. 4.32 Installing a dynamometer (*Source:* Leutert document, Composite Catalog, 1990–1991).

B. Interpreting representative diagrams (circular diagrams involving 24 hours of recording time)

a. Normal operation

1) Under normal operating conditions, the ammeter recorder should plot a curve with an axial symmetry and an amperage value equal to or close to the value indicated on the nameplate. Figure 4.33 illustrates this ideal condition.

2) As long as the curve is symmetrical and constant the pump is operating properly. Any deviation from this normal diagram is due to abnormal operation of the pump or changes in well conditions.

b. Overloading

Figure 4.34 shows the diagram of a unit that has stopped because of an intensity overload. Section A and section B of the curve show start up followed by operation with normal amperage. Section C shows a gradual increase in amperage until it goes over the limit, then shutdown because of an intensity overload.

Note: Do not attempt to start up again until the cause of the overload has been corrected.

Usual causes of shutdown:

- increased specific gravity of the pumped fluid and as a result increased power and intensity

- sand production
- emulsion or increasing viscosity
- mechanical or electrical problems such as overheating of the motor or equipment fatigue
- miscellaneous electrical problems.

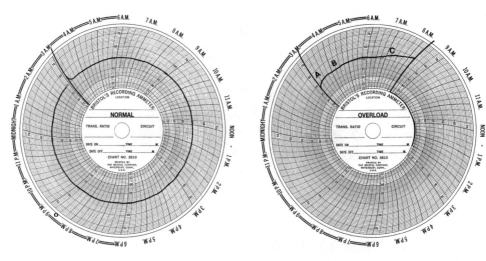

Fig. 4.33 Normal diagram.

Fig. 4.34 Diagram with overloading.

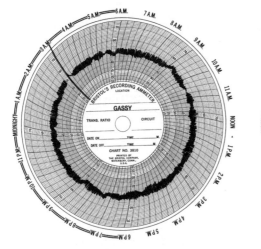

Fig. 4.35 Diagram with a gassy fluid.

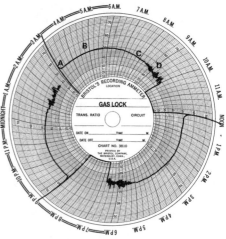

Fig. 4.36 Diagram with gas lock.

(*Source:* After Reda documents).

4. ARTIFICIAL LIFT

c. Pumping a slightly gassy fluid

Figure 4.35 shows the diagram of a unit that is properly rated but is pumping a slightly gassy fluid. The variations in amperage are caused by pumping liquid and gas with different specific gravity intermittently. This condition is often seen in conjunction with a reduction in production.

d. Gas lock in the pump

Figure 4.36 shows an amperage diagram for a pump that is gassed up and continues to operate with a slightly lower amperage. If the amperage drops significantly the pump may stop.

1) Section A: The pump starts up. At this time the fluid level in the annulus is high, so the flow rate is higher (due to a lower delivery head requirement) and therefore the amperage is higher than normal.
2) Section B: The curve is normal as long as the liquid level remains within specified limits.
3) Section C: The amperage drops because the liquid level falls below specified conditions. Then there is a variation due to the gas beginning to be released at the pump inlet.
4) Section D shows disturbed amperage due to gas interfering because of a liquid level near the pump inlet. If the gas lock continues to increase the amperage may drop and trigger shutdown of the pump.

4.1.7 Defining a pumping installation

Whatever the type of pumping, and as mentioned in Section 4.1.2.3 for rod pumping, the first steps are as follows depending on reservoir characteristics:

- Choose the required flow rate.
- Determine the pump's minimum corresponding depth with due consideration to the dynamic level (be careful with the variation in dynamic level versus the declining reservoir pressure) and the required submergence (to reduce the amount of free gas at the pump intake).

With respect to the pump's real depth, note that:

- In rod pumping, the pump is not usually placed much lower in order to limit stress on the rods.
- In the other cases, setting the pump in the "bottom" of the well can be contemplated from the start.

4.1.7.1 A sucker rod pumping system

A. General introduction

When a rod pumping system is being defined, simplifications are employed which can be broken down into two categories in particular for the equipment in the well:

- a simplified representation of the sucker rod string, such as for example the hypothesis in some methods that the mass of the sucker rod string is concentrated in one point
- a simplified consideration of the action of the downhole pump, particularly with respect to valves opening and closing.

Consequently, the formulas may yield incorrect results especially for deep wells or for high pumping rates. Care is also required for deviated wells, or when the well profile is not regular, the fluid is very viscous, or excessive sand or gas is produced through the pump, etc.

The final determination is reached by successive approaches. Three stages can be distinguished in the basic process:

a) An initial selection of operating parameters and components making up the system, with the main information that has to be known or supposed as follows:
  - the liquid level
  - the pump depth
  - the pumping rate
  - the surface stroke length (polished rod stroke)
  - the pump and plunger diameter
  - the fluid density
  - the tubing nominal diameter
  - whether the tubing is anchored or free
  - the composition of the pumping string (particularly the sucker rod diameter).

b) Based on the above, an attempt is made to determine the corresponding operating parameters with an appropriate method, in particular:
  - for the downhole equipment:
    - the plunger stroke
    - the volume effectively pumped
    - the peak load on the polished rod
    - the minimum load on the polished rod
  - for the surface unit:
    - the maximum reducer torque
    - the power required at the polished rod
    - the required counterweight
    - the power rating for the motor installed on the unit.

c) The operating parameters that have been determined in this way are then compared with the required flow rate and with the limitations specific to the installation which was previously selected.

Generally speaking it will be necessary to repeat the selection process several times before getting results consistent with the chosen equipment and optimizing the selection.

B. The main methods

The main methods that are or have been used in defining a rod pumping system are the following:

a) The "traditional" method using the formulas worked out by J.C. Slonneger, Mills, Coberly, etc. based principally on stress in the sucker rods. Though it is less commonly used now, we will go into detail on the main stages of the method in Section 4.1.7.1 C. because it illustrates clearly:

- the process of defining the system by successive approaches
- how the different parameters to be determined are interdependent.

b) The "API" method based on work done by the Sucker Rod Pumping Research, SRPR, between 1959 and 1965 and focusing more on calculating the maximum torque and the counterweight. It is developed by API in API RP 11L (RP: recommended practice). In particular the method uses correlations between a number of non-dimensional variables. The correlations were proposed by the SRPR after analyzing a great many dynamometric recordings obtained from an electric analog simulator. The use of non-dimensional parameters allows the problem to be generalized.

Among the non-dimensional variables, the two main ones are:

1) a non-dimensional pumping speed corresponding to the ratio between the pumping rate and the natural frequency of the sucker rod string;

2) a non-dimensional elongation of the polished rod corresponding to the ratio obtained by dividing the difference in force (due to the fluid load on the plunger) between the upstroke and the downstroke (see F_o on the simplified dynamometric diagram in Diagram 4.1) by the force required to elongate the whole rod string by a length equal to the polished rod's stroke.

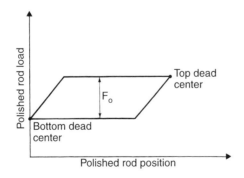

Diagram 4.1 (*Source:* ENSPM Formation Industrie).

Today the major rod pumping equipment manufacturers propose readily implemented PC programs using the API method.

c) The "wave equations" method that uses differential equations and is mainly employed to:
- finalize the choice between several alternative configurations predetermined with the API method
- determine particular facilities, for example if fiberglass sucker rods are used (the other methods can not be applied to this case)
- make a more refined analysis of the measurements on a well in operation.

The method requires the use of a PC program and a longer processing time than the computerized API method.

C. Presentation of the traditional method

It should be born in mind that the choices and determinations made in each of the stages described below may be incompatible with those made previously. As a result, compatibility must be checked at each stage and if need be the previous choices must be modified accordingly. Then the stages downstream from the modification must be reviewed.

a. Plunger and tubing diameter

The plunger diameter is chosen mainly depending on the required flow rate and then on the pump depth. Charts and tables based on experience can make the choice easier and limit the number of iterations that later stages will entail.

b. Pumping rate and stroke

For a given stroke, the pumping rate required to obtain the target flow rate can be calculated (or vice versa). The selected stroke length will of course have to be given due consideration when the surface unit is chosen in stage e). In the present stage, the surface unit can even be predetermined (by means of charts or tables based on past experience) mainly versus the plunger diameter, the pump depth and the pumping rate.

c. Sucker rod string

The string is chosen according to the plunger diameter, the pump depth and the pumping rate. The pumping rate must be checked to be sure it is not synchronous with the natural frequency of the rod string, given the type of string and the pump depth. Charts and tables based on past experience make this choice easier.

d. Peak and minimum load on the polished rod

These are calculated on the basis of the weight of the sucker rod string in air and the fluid load on the plunger with appropriate consideration given to an impetus factor (a dynamic factor depending on the pumping rate and stroke) and buoyancy.

Based on the peak load, the maximum stress is calculated and then compared with the allowable stress, which is dependent on the rod string that was previously chosen. The peak load also corresponds to the required pumping beam capacity.

4. ARTIFICIAL LIFT

The load factor is also calculated, it corresponds to the ratio between the minimum and peak loads. Along with the peak load it is a key parameter in studying sucker rod fatigue.

e. Counterweight, maximum torque at the reducer outlet and surface unit

The counterweight is calculated on the basis of the peak and minimum load. The maximum torque at the reducer outlet is calculated using:
- the difference between the peak load and the counterweight
- the stroke.

Then the surface unit can be chosen based on the rated or selected:
- maximum torque at the reducer outlet
- peak polished rod load
- and polished rod stroke.

The choice will have to be checked to be sure it is coherent with the allowable excess in reducer load, with the minimum and maximum counterweights specific to the selected surface unit.

f. Actual flow rate

The first calculation is the sucker rod elongation (and that of the tubing if it is not anchored) due to the fluid load being transferred alternately from the tubing to the sucker rods and vice versa.

The plunger overtravel due to the dynamic effect is also calculated. The downhole plunger's actual stroke is deduced from the above (in practice it is shorter than the polished rod's stroke because of elongation despite the overtravel effect), then the actual flow rate according to the pumping rate and the pump's volumetric efficiency.

g. Motor and power rating

Given the parameters determined in the previous stages and depending on the different volumetric and mechanical efficiency ratios, the power required at the polished rod and at the motor can now be calculated.

The motor can then be chosen.

4.1.7.2 Submerged centrifugal pumping

First of all a reminder that these pumps are sensitive to high viscosity and to excessive amounts of free gas present at the pump intake. Defining the system involves the following main stages:

a. Amount of free gas at the pump intake

This is determined on the basis of:
- the characteristics of the effluent (and therefore the bubble point pressure, dissolved gas at the bubble point and the specific gravity of the oil)

- the pressure and temperature at the pump intake (and therefore the reservoir characteristics, the required flow rate and the pump depth).

b. Whether to use a gas anchor

The decision is made according to the ratio between the free gas flow rate at the pump and the overall flow rate (oil + free gas + water) all under bottomhole conditions. A gas anchor is recommended if the ratio is greater than 10%.

c. Characteristic curve (required total dynamic head, TDH, versus flow rate) in the target flow rate range

In order to plot the curve the following points in particular are taken into consideration:
- the hydrostatic pressure and pressure losses in the tubing (and therefore the pump depth, the specific gravity and viscosity of the effluent)
- the available suction pressure and the pressure required at the wellhead.

d. Pump and the number of corresponding stages

According to the total dynamic head which is required, a pump suited to the casing (available space) and to the required flow rate (appropriate efficiency) is found in manufacturers literature, along with the number of stages necessary (based on the delivery head per stage).

e. Motor, voltage and intensity

Manufacturers literature is scanned to find a motor suited to the casing (space available) taking the required power rating into consideration. This is determined from the power necessary per stage, indicated by the manufacturer, corrected for the fluid's specific gravity and the number of stages.

The fluid's velocity as it goes by the motor is checked to be sure it is at least equal to 30 cm/s (1 ft/s) and imperatively greater than 6 cm/s (0.2 ft/s).

Then the required voltage (as high as possible) and intensity (as high as possible) are determined that are compatible with the motor and the required power rating. In fact, when a standard pump is used the choice of the pump determines the motor that goes with it.

f. Cable

The choice depends on the temperature (with respect to the type of cable) and the voltage drop in the cable (with respect to its diameter). The voltage drop is determined by using a correction that depends on the temperature and must be lower than both 10 V/100 m (330 ft) and 12.5% of the total voltage.

g. Surface electric equipment (transformer)

This choice depends on the power rating of the motor and on the required voltage on the surface.

4.1.7.3 Hydraulic plunger pumping

Before making the calculations on the system, the type of circuit and the power fluid must be chosen (see Section 4.1.4). Then the main stages are as follows:

a. Pump

The pump is chosen mainly according to the flow rate that will be required and the space available. The ratio between the driving plunger and the pump plunger cross-section in particular and their travel are taken into account.

b. Flow rates (injection and return) and pumping speed

The injection rate is obtained from the required flow rate, the total pump efficiency (including the efficiency due to the presence of gas if applicable), and the ratio of the plunger (driving and pump) cross-sections.

The return rate is equal to the required flow rate plus the injection rate.

The pumping speed is deduced from the injection rate taking the motor's efficiency and travel into account.

c. Pressure head of the surface injection pump

To determine the pressure head, the following points need to be assessed according to the pump depth and the viscosity and density of the fluids:

- the pressure losses on the surface between the pump and the wellhead, in the injection tubing, in the pump, and in the return circuit up to the wellhead
- the hydrostatic pressure of the column of working fluid and the column of return fluid
- the specified pressure at the beginning of the flowline.

d. Power rating and surface unit

The required power is deduced using the necessary pressure and the injection rate, according to the surface pump efficiency. The surface unit can then be chosen.

It may prove necessary at this stage, as well as at the previous one, to proceed to an iteration with a new choice of pump and/or any new base data (among which: specified flow rate, pump depth, type of circuit).

4.1.7.4 Hydraulic jet pumping

Here again the first thing to be done is to choose the type of circuit and the power fluid. The main stages in determining the system are as follows:

a. The pump (ejector-diffuser)

This choice is made after having determined the minimum cross-section (to prevent cavitation) of the ejector-diffuser throat annulus, and this depends on the specified flow rate and on the submergence depth of the pump.

b. Injection pressure

This stage requires a fairly complex calculation phase by iteration. Based on an assumed likely injection pressure (P_{inj}) (15 to 30 MPa, or 2000 to 4000 psi) and the specific gravity of the injected fluid, the following are calculated:

- the injection flow rate based on the ejector pressure (P_{eject}) and therefore on the pressure loss in the injection tubing (but these two terms depend on the injection flow rate), on the submergence pressure (P_{subm}), on the specific gravity of the injected fluid and on the cross-section of the ejector
- the return flow rate as a direct result of the target flow rate and the injection flow rate
- the diffuser pressure (P_{dif}), a function of the hydrostatic pressure of the return fluid (and therefore of its specific gravity), of the pressure loss in the return circuit and the specified pressure at the beginning of the flowline
- the ratio (N) of the pressure variations $[(P_{dif} - P_{subm})/(P_{eject} - P_{dif})]$ versus the ratio of the suction and output mass flow rates and the ratio of the ejector and annulus cross-sections.

Based on this ratio N, an ejector pressure then a surface injection pressure are recalculated which are compared with the assumed pressure. Then this whole calculation stage is reiterated until the two figures are the same and satisfactory. This may entail having to choose another ejector-diffuser couple and/or modifying the base data.

c. Maximum flow rate without cavitation

Depending on the effective annulus cross-section, the maximum flow rate without cavitation can be determined.

d. Power rating and surface unit

The power rating is deduced from the surface injection pressure, the pressure loss between the surface pump and the well, the specified flow rate and the surface pump efficiency. The surface unit can then be chosen.

4.2 GAS LIFT

4.2.1 Principle and types of gas lift

4.2.1.1 Principle

Gas lift is an artificial lift production technique that is used to make a non-flowing or insufficiently flowing well come on stream by lessening the hydrostatic back pressure between the bottom of the hole and the surface. This is done by injecting makeup gas at the base of the production string.

4.2.1.2 Types of gas lift (Fig. 4.37)

A. According to injection method

The gas can be injected either continuously or intermittently.

Continuous gas lift: natural gas is injected at a given pressure and flow rate at the base of the production string. This makes the density of the fluid in the tubing lighter and allows the mixture of the two fluids to rise up to the surface, thereby making the well flow again.

Intermittent gas lift: a given volume of pressurized gas is injected at a high flow rate at the lower part of the production string. This flushes the volume of liquid it contains upward. As a result, the pressure on the pay zone decreases and it starts flowing again. The liquid accumulating above the injection point will be flushed out in the same way.

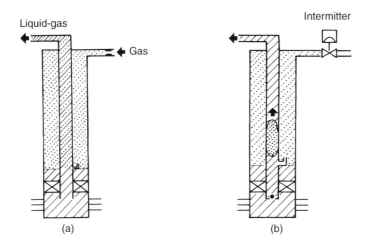

Fig. 4.37 Types of gas lift (*Source:* ENSPM Formation Industrie).
a. Continuous gas lift. b. Intermittent gas lift.

Limitations of continuous and intermittent gas lift

- The continuous method is usually suited to wells with a good productivity index ($PI > 10$ m^3/d/MPa, or 0.45 bbl/d/psi) and the intermittent version to wells with a low productivity index ($PI < 10$ m^3/d/MPa, or 0.45 bbl/d/psi). In practice it can be seen that:

- continuous gas lift is suited to a liquid output of 30 to 3000 m^3/d (200 to 20 000 bpd)
- intermittent gas lift is used for outputs of less than 80 m^3/d (500 bpd)
- in the overlap between 30 and 80 m^3/day (200 and 500 bpd), a low *PI* well will preferably be produced with a smaller production string diameter to the benefit of increased upward fluid velocity
- the efficiency of intermittent gas lift is much lower than that of the continuous method because the energy of the compressed gas under the liquid slug is lost when the gas gets to the surface
- the intermittent method is the only gas lift technique possible in good producers that have a low downhole pressure because the reservoir pressure is initially low or has been depleted
- in summary, 95% of gas lifted wells produce by continuous gas lift.

B. *According to the surface injection circuit*

The gas used often comes from the formation GOR (gas/oil ratio) of the relevant oil reservoir or from nearby available gas wells. Two injection circuits can then be contemplated:

- closed circuit, and
- open circuit gas lift.

  a) Closed circuit means that the gas that has been used is recovered at the separator outlet. After undergoing various possible processing stages (gasoline extraction, dehydration, sweetening, for example), it is recompressed by a series of compressors and reinjected into the wells (Fig. 4.38).

  b) Open circuit involves processed gas from a gas reservoir and is flared or marketed after it is used.

Note: Wells may produce oil by self gas lift. As shown in the simplified diagram (Fig. 4.39), the oil from the oil reservoir is lifted by means of the gas from a gas reservoir located above it. The gas enters the production string through perforations and an injection device located between two packers (this is illustrated by Askarene and El Adeb Larache in Algeria).

C. *According to the type of completion*

Gas lift can be used with single- or multiple-zone completion and well production can be:

- direct, i.e. injection through the casing, production through the tubing; or
- reverse, i.e. injection by the tubing with production through the casing.

Additionally, macaronis allow gas lift implementation by concentric completion.

4. ARTIFICIAL LIFT

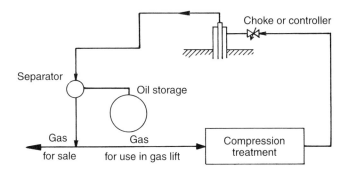

Fig. 4.38 Surface facilities for closed circuit gas lift (*Source:* ENSPM Formation Industrie).

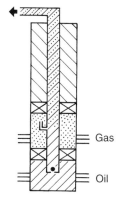

Fig. 4.39 Self gas lift (*Source:* ENSPM Formation Industrie).

4.2.2 Regarding the well

4.2.2.1 Operating conditions in continuous gas lift

What must in particular be determined is the depth of the injection point, the injection pressure and flow rate. These three parameters are interrelated. For a given flow rate, the deeper the injection point, the less gas needs to be injected but the higher the required injection pressure.

4. ARTIFICIAL LIFT

The final decision must take the pressure and flow rate conditions of the available gas into consideration as well as the cost of any recompression. The economics of the system is highly sensitive to the wellhead pressure and consequently to the separation pressure. The separation pressure should be kept as low as possible given the separation and recompression requirements (for closed circuit gas lift).

Figure 4.40 illustrates how the operating parameters for a well producing by gas lift are determined on a graph, where:

- H vertical depth of the zone in the well that is brought on stream
- P_{WH} pressure required at the wellhead
- P_R reservoir pressure
- Q specified liquid flow rate
- PI liquid productivity index
- P_{BH} bottomhole pressure at flow rate Q
- G_N average flowing pressure gradient below the injection point (depends on the amount of naturally produced gas)
- $G_{(N+I)}$ average flowing pressure gradient above the injection point (depends on the amount of naturally produced gas and the amount of injected gas).

The graph construction in Fig. 4.40 shows how the following parameters are determined:

- H_I depth of the injection point
- P_I gas injection pressure required at the casing head
- Q_I injection gas flow rate:
    taken from: $Q_I = Q_{(N+I)} - Q_{(N)}$
    with: $Q_{(N+I)}$ gas flow rate corresponding to $G_{(N+I)}$
             $Q_{(N)}$ gas flow rate corresponding to $G_{(N)}$.

In an open circuit system, the maximum possible injection depth and minimum corresponding required injection flow rate are determined based on the available injection pressure.

In a closed circuit, the required injection pressure and the corresponding injection flow rate are determined based on an injection depth. The power rating and therefore the cost of recompression can be deduced from them. Based on another depth the same can be done and in this way the system can be optimized with respect to operating costs.

Optimization is no easy job since the parameters involved (P_R, Q, WOR: water oil ratio) vary during the reservoir's productive life. In the intermediate period (a reservoir that is starting its depletion phase or when depletion is underway) when the bottomhole pressure is still much higher than the pressure used in determining the final injection depth, the pressure required in order to inject at this depth is higher (see point B Fig. 4.41). Here it is preferable to inject the gas at a higher point (see point C, Fig. 4.41).

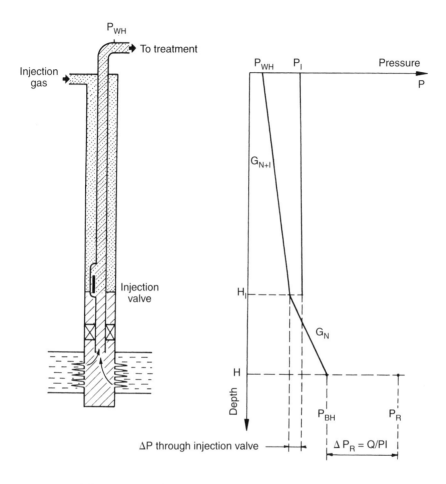

Fig. 4.40 Graph showing determination of a continuous gas lift system (*Source:* ENSPM Formation Industrie).

4.2.2.2 Pressure gradients in producing wells

The aim of gas lift is to reduce the pressure drop in the tubing when the reservoir is being produced. The pressure drop is due to:

- the weight of the fluid on its way to the surface
- losses by friction caused by fluid flow.

It is complex to calculate the pressure gradients because there is a two-phase flow between the bottom and the surface with expanding gas in vertical or inclined pipe. There has been a large body of research on two-phase flow in conjunction with laboratory and field testing.

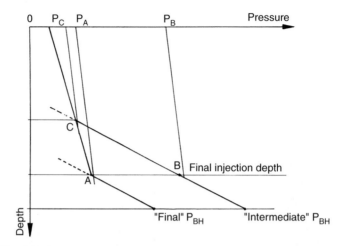

Fig. 4.41 Determining the optimum injection point according to the change in bottomhole pressure (*Source:* ENSPM Formation Industrie).
P_A final injection pressure,
P_B injection pressure required with "intermediate" P_{BH} and injection at final injection depth,
P_C injection pressure required with "intermediate" P_{BH} and optimized injection depth.

They have been used to adapt theoretical calculations of pressure variations to real cases in production tubings. The conclusions of these studies are presented in the form of pressure gradients in particular.

The correlations involve the variation in pressure with depth versus variations in GLR (gas/liquid ratio), WOR and the liquid's density for different tubing diameters. The GLR is the ratio of an amount of gas over a given amount of liquid (oil + water). It is expressed in cubic feet per barrel or in cubic meters per cubic meter under standard conditions.

Note that even though the GLR is a volume ratio, since the American units used to measure a gas flow rate (cubic feet) or an oil flow rate (barrels) are different, they will make the GLR expressed in American units different from the GLR in metric units (1 m^3/m^3 is approximately equal to 5.6 cu ft/bbl).

It is indispensable to use the gradients to solve gas lift problems and they can be:

- applied in a computer program, e.g.:
  - DIFA program (EAP) based on research by Dun and Ros
  - Camco program after Poettmann and Carpenter
  - "Pépite Puits" program based on joint Elf, Total and IFP research
- gathered together in the form of a collection of curves to be used in solving gas lift problems by the graph method (Fig. 4.42).

4. ARTIFICIAL LIFT

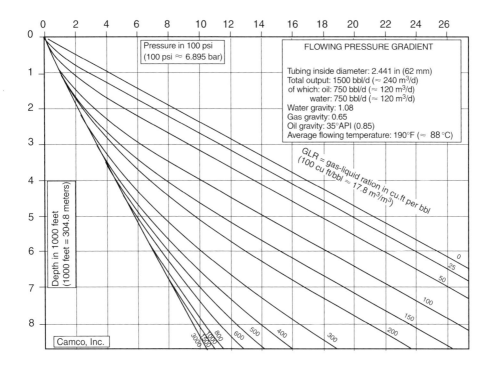

Fig. 4.42 Example of pressure curves for a producing well (*Source:* After a Camco document).

In the second instance the curves are presented as described in the following section.

For a crude, a gas, water of given characteristics, for each inside diameter of tubing, for various liquid flow rates, BSW and bottomhole temperatures, the variation in the vertical back pressure is given according to a suite of GLR curves ranging from zero GLR (no gas) to an optimum GLR value. The optimum value corresponds to a minimum gradient and therefore represents the maximum amount of gas to be injected. Any further injection of gas would introduce more pressure losses by friction than the gain in making the fluid column lighter.

The curves are convenient and quick to use. The collections supplied by manufacturers are complete enough so that the real problem data can be adjusted with sufficient accuracy to the theoretical data used to establish the curves. They are a must for verifications or for dealing with special isolated cases.

4.2.2.3 Unloading a well at start up with unloading valves

A. Need for extra injection points

When starting up a well with gas lift, the available gas injection pressure is much lower than the hydrostatic pressure generated at the injection depth by the height of "dead" fluid in the well (Fig. 4.43). This is true with the exception of some very special cases of shallow wells with a ready source of high pressure gas. The high-gradient dead fluid may be gas-free effluent or completion fluid. It will therefore be necessary, for both types of gas lift (continuous and intermittent), to place extra injection points in the column. They are designed to clean out the annulus and unload the dead fluid gradually in phases by gassing up the column in stages. The aim is to make the effluent lighter little by little until the gas can go through the final injection port.

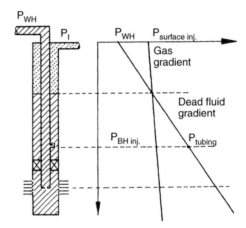

Fig. 4.43 Need for unloading valves (*Source:* ENSPM Formation Industrie).

B. Need for unloading valves

The dead fluid could be unloaded by perforating the tubing but this would result in excessive gas consumption and a limited final injection depth since the gas going through the upper ports would be lost for the deepest one. This is why the production string is equipped with unloading valves. The principle is the same as for a pressure regulating valve that opens as soon as the gas pressure goes over a threshold that the valve is preset for. Likewise, the valve closes when the gas pressure drops below the predetermined value.

The valves are placed along the tubing string and locked in special mandrels at predetermined depths. The depths depend on the available injection gas pressure at the wellhead, the gas gradient, the static gradient of the dead fluid and the tubing back pressure at the flow rate given by the unloading GLR. However, the uppermost valve is usually placed and set for a temporary overpressure, the kick-off pressure. The valves are set for closing pressures that decrease the deeper they are located.

This allows the following to be achieved automatically without having to modify anything on the surface:
- the well is unloaded down to the final injection point
- the unloading valves close as the annulus is cleared out.

The casing head pressure drops automatically because:
- there is less pressure loss with gas than with liquid through the new valve that has just emerged
- the back pressure in the tubing falls as the tubing effluent is gassed
- the restriction to gas flow on the surface is greater than through any of the gas-lift valves in the well.

C. Simplified analysis of start up with continuous gas lift

In the beginning the valves are closed except for the ones that are located deep enough for the hydrostatic pressure of the liquid to be higher than their opening pressure.

As soon as pressure is exerted on the annulus all the valves open (Fig. 4.44a). The uppermost valve is submerged. Pushed by the injection gas pressure, the annular liquid is transferred into the tubing by the principle of communicating vessels through the open valves. Then the first valve emerges. The gas enters the tubing and ejects the liquid that is in it. The level keeps on getting lower in the annulus (Fig. 4.44b) and the liquid displaced from the annulus into the tubing is produced gradually by gas lift from the first valve.

The process continues until the time when the second valve emerges (Fig. 4.44c). Then the liquid slug between valves 1 and 2 is ejected from the level of the second valve. Since the opening pressure for valve 2 is lower than for valve 1, the gas pressure in the annulus decreases and valve 1 closes (Fig. 4.44d). Then as the level continues to fall the liquid displaced from the annulus into the tubing is produced on the surface by gas lift from the second valve.

The process takes place until the third valve emerges (Fig. 4.44e). Then the liquid slug between valves 2 and 3 is ejected from the level of the third valve, since valves 1 and 2 are closed.

In the same way the process continues until the final injection depth is reached (Fig. 4.44f), with the annulus gradually emptied and the injection point reached without any need for a source of high pressure gas.

This well unloading operation is automatic but still requires an operator to be present. If need be he will monitor the changes in annulus level with an echometer. Note that until the first valve opens the injection flow rate must be held down so that the pressure increase in the annulus is progressive and the velocity of the liquid through the valves is not too high. If the liquid velocity was too high they might be damaged by erosion.

4. ARTIFICIAL LIFT

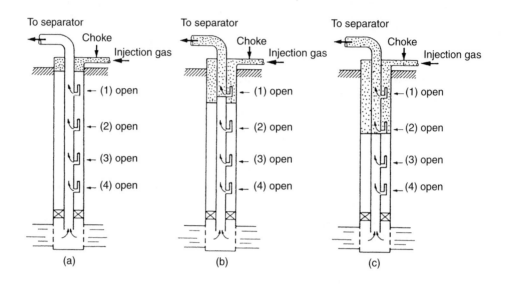

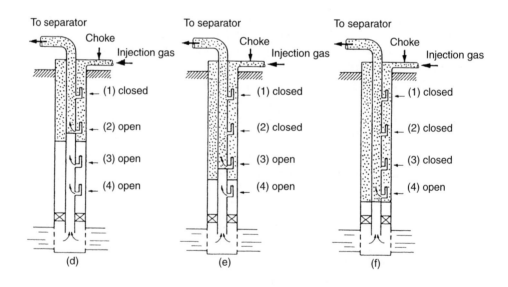

Fig. 4.44 Simplified analysis of start up with continuous gas lift (*Source:* After a Camco document).

4.2.2.4 Gas-lift valve technology

A. General description (Fig. 4.45)

Gas-lift valves are gas nozzles that work like on/off control valves. A valve consists of two parts:

- the valve itself
- the latch to set the valve in the valve mandrel.

General principle of a gas-lift valve

The body of the valve is made of stainless steel. Inside it a hemispherical tungsten carbide poppet valve opens or closes on a tungsten carbide or Monel seat called a port and sized according to the requisite flow rate. The poppet valve is connected by a stem to the servomotor. Most gas-lift valves have check valves at the base of the body which have practically no resistance to the gas passing and prevent the reverse return of the fluid, thereby keeping the annulus from getting filled up when the well is shut in. They are also very useful in the event of further stimulation operations at a later date.

The servomotor that allows the poppet valve to open or close on its seat can be of two types:

- pneumatic with a bellows
- mechanical with a spring.

B. Bellows pneumatic valve (bellows valve)

The servomotor is a bellows chamber full of pressurized nitrogen and preset by calibration via an admission or decompression valve located on the upper part of the chamber. The bellows consists of 2 or 3 concentric Monel tubes of only slightly different diameters and a thickness of 5/1000". The tubes are cold drawn and bent. The bellows is silver soldered to the chamber (on some valves a silicon liquid partially fills the bellows and passes through a port to damp valve movements). The compression of the bellows represents the opening stroke for the valve.

The valves operate through an opposition between forces resulting from:

- the pressure of the injected gas and the fluids in the tubing at valve depth, and
- the pressure of nitrogen in the bellows that may be enhanced by a spring.

The valve is adjusted by choosing the nitrogen pressure in the bellows. When it is calibrated on the surface to adjust the nitrogen pressure, a temperature correction coefficient must be used to account for the difference between downhole and test rack conditions.

Example: Principle on which a casing operated bellows valve works. This valve (Fig. 4.46) will work as an upstream pressure regulator. It opens when the force lifting the poppet valve up exceeds the force holding it down on the seat:

$$P_{cgs}(A_b - A_v) + P_{tbg} \times A_v > P_b \times A_b$$

The valve closes when:

$$P_{cgs} \times A_b < P_b \times A_b$$

i.e.:
$$P_{cgs} < P_b$$

where:

- P_b pressure in the bellows
- A_b area of the bellows
- A_v area of the valve
- P_{csg} casing (annulus) pressure
- P_{tbg} tubing pressure.

Opening is therefore governed by tubing and injection pressure while closing is controlled by injection pressure alone.

C. Spring loaded valve (spring valve)

The poppet valve and its stem are connected to a bellows with no inside pressure (atmospheric pressure equivalent to zero) that only serves to transmit forces and is not influenced by temperature. Downhole the opening and closing pressures are controlled by the action of a calibrated spring.

In order to adjust the valve the compression force is turned into a pressure effect. Because there is moreover no temperature effect, calibration on the surface is very simple since one turn of the screw on the spring's stop represents a specified pressure.

Figure 4.47 illustrates a casing operated type valve: the pressure in the annulus acts on the bellows in opposition to the spring.

Figure 4.48 illustrates another type of valve. The annulus pressure or P_{csg} acts to open the valve only on the area of the valve while the tubing pressure is exerted on the force transmitting bellows. Consequently the casing pressure effect is much less important than the tubing effect for opening. In addition, since the valve orifice offers a restriction to the fluid, only the opposition between P_{tbg} and the spring acts to close the valve. The valve is termed production pressure operated by manufacturers and is often called tubing operated by users.

D. Comparison and use of the two types of valves

Without going into great detail on the advantages and drawbacks of the two types of valves, we would like to point out the main observations about their use as such.

a. Bellows valve

It has the qualities of pneumatic control: the response is precise, flexible and very sensitive. Surface calibration is fine tuned (in particular the temperature effect on the nitrogen pressure is corrected) and virtually trouble-free operation can be expected once in the well.

4. ARTIFICIAL LIFT

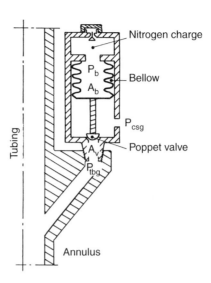

Fig. 4.46 Simplified diagram of a casing operated valve (*Source:* ENSPM Formation Industrie).

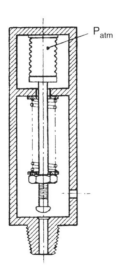

Fig. 4.47 Casing operated type spring valve (*Source:* ENSPM Formation Industrie).

Fig. 4.45 gas-lift valve (*Source:* McMurry).

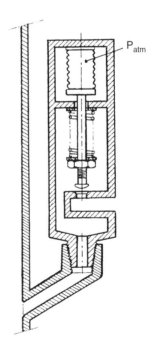

Fig. 4.48 Tubing operated type spring valve (*Source:* ENSPM Formation Industrie).

This type of valve today equips over 90% of the gas lift systems. It is particularly well suited to gas lift wells with casing operated valves where the temperature of the injection gas is practically stable and holds no surprises.

b. Spring valve

It works less flexibly than the air actuated valve, since the spring's response is not very sensitive and its mechanical characteristics are affected with time. In contrast, it is not influenced by temperature and this not inconsiderable advantage means that it can be used in a tubing operated system. In fact a well's production and consequently the fluid's temperature are not necessarily stable. Since it is production that governs the opening and closing of a tubing operated valve, it must not be dependent on the effect of temperature.

Generally speaking, a tubing operated spring is suited to wells producing by parallel dual tubing string completion:

- a common source of gas in the annulus supplies two tubing strings that produce fluids that may have widely different characteristics (pressure, density, GLR, etc.)
- to make it easier to control the whole system it is usually preferable to equip one of the tubings with casing operated valves and the other with tubing operated.

4.2.2.5 Tubing equipment specific to gas lift

A. Valve mandrels

There are three main types of mandrels:

- what are termed conventional mandrels
- side pocket mandrels
- mandrels for concentric valves.

a. Conventional mandrels (Fig. 4.49)

They are made from sections of tubing. The valves and check valves are screwed on before they are run in at the base of a ported receptacle, which allows the gas to flow from the annulus to the tubing. The valve is vertically positioned with the help of a small guide attached to the mandrel body which also serves as a protector during running and pulling operations. This equipment was the only type used before the side pocket mandrel was invented and is mainly employed in the United States on shallow onshore wells where workover costs are not prohibitive. It is a fact that any trouble with a valve means that the well has to be killed and the equipment pulled out.

b. Side pocket mandrels (Figs. 4.50 to 4.53)

Introduced by the American manufacturer Camco in 1954, side pocket mandrels in conjunction with wireline operations have revolutionized gas lift technology. The mandrel, made of forged steel, is an asymmetrical tube with an aperture in it. A pocket is connected to the aperture and has a machined valve seat at its base where there are holes communicating with the annulus. The two ends have standard tubing threads.

Fig. 4.49 Conventional mandrel (*Source:* Camco).

ture and has a machined valve seat at its base where there are holes communicating with the annulus. The two ends have standard tubing threads.

Incorporated in the tubing string (Fig. 4.51), the mandrels are run empty or with dummy valves (blind valves used to test the tubing) during equipment installation. Then the gas-lift valves are set by wireline operations. The valves are equipped with latches that keep them in place in the seats. In addition, packings provide a seal on either side of the injection point (Fig. 4.52).

The wireline technique that uses specific equipment to set and retrieve valves at any time without killing the well is totally operational and can even be used in highly deviated wells.

Lastly, side pocket mandrels also allow the injection point depth to be adapted as the bottomhole pressure declines. Figure 4.53 is an example of a widely used mandrel with its main dimensional data.

c. Mandrels with concentric valves

There are also special mandrels on the market without side pockets, but equipped with concentric valves. The gas flows from the annulus into the concentric valve by ports similar to the ones in circulating sleeves.

4. ARTIFICIAL LIFT

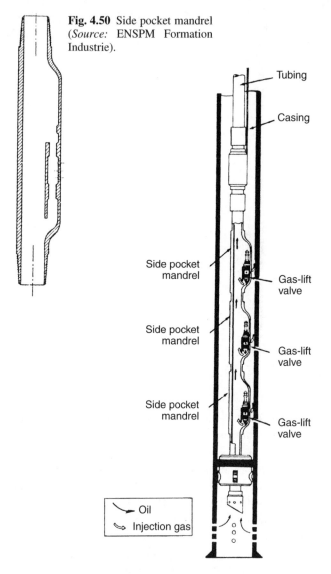

Fig. 4.50 Side pocket mandrel (*Source:* ENSPM Formation Industrie).

Fig. 4.51 Tubing equipped with side pocket mandrels (*Source:* After a Camco document, Composite Catalog 1982–1983).

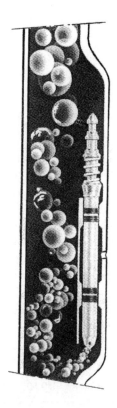

Fig. 4.52 Gas-lift valve in a side pocket mandrel (*Source:* Camco).

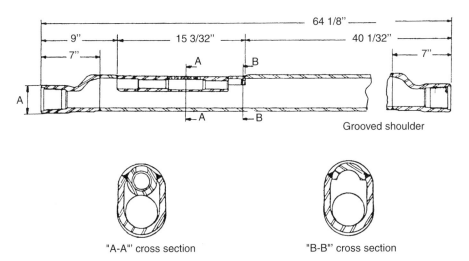

Fig. 4.53 Cross-section of a KBM type mandrel (*Source:* J. de Saint-Palais, J. Franc: *Cours de gas lift,* Éditions Technip, Paris, 1969).

The concentric valve can:
- either be integrated in the mandrel and the tubing must be pulled to change the valve
- positioned by wireline and here the valve must be pulled up to the surface if access is required deeper by wireline.

Mandrels with concentric valves can solve a number of problems such as space available in multiple completion, etc.

B. Other specific equipment

In addition to the packer, specific equipment can be used such as:
- check valve
- annulus safety valve
- tubing-head spool.

a. Check valve

A check valve can be placed at the base of the tubing to prevent any possible fluid return into the formation. It is indispensable in intermittent gas lift when the reservoir is seriously depleted.

b. Annulus safety valve

In addition to the conventional tubing safety system, a gas lifted well can be equipped with a further annulus safety system (subsurface tubing-annulus safety valve). It is in fact a packer (see Chapter 3, Section 3.6.2.4) located in the upper part of the well. It features a by-pass for the injection gas controlled by a control line and allows the volume of gas to be kept in the annulus in the event of an accident (wellhead failure, fire, etc.). It is mainly justified offshore where a large pressurized annular volume is a potential danger for the environment.

c. Tubing-head spool (Fig. 4.54)

It is necessary to be sure that the pressurized annulus can not endanger the last intermediate casing. This can be done by equipping the base of the tubing-head spool with an X bushing type isolating seal for example.

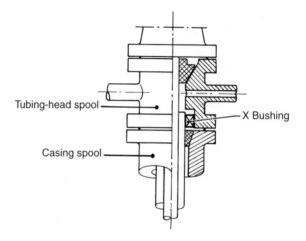

Fig. 4.54 Tubing-head spool (*Source:* ENSPM Formation Industrie).

4.2.2.6 Kickover or positioning tools

The technology and basic equipment for wireline work are discussed in detail in Chapter 5 (Well Servicing and Workover) in Section 5.2.1. Here we are only considering the tool that allows access to the side pockets.

In the wireline string between the valve running/pulling tool and the jar, a special kickover tool is inserted. It is used to set the valve or to retrieve it in its side pocket. The kickover tool is usually found in two different versions:

- with extendable arms
- with pivoting arms.

The examples below show how each type of tool works.

4. ARTIFICIAL LIFT

a. Extendable arm kickover tool:
 setting a valve with Camco's L kickover tool (Fig. 4.55)

The tool is connected to the running string by a knuckle joint and is used to position and lock a valve in a side pocket mandrel. It has two extendable centering arms fixed onto a sliding system. In the raised position the centering arms are kept closed by a sliding bushing that remains locked as long as the string is in a running in position. Pulling the string back out unlocks it and allows the arms to extend. Then the tool tilts over in the space provided in the mandrel and its axis lines up with the axis of the side pocket (Fig. 4.56).

b. Pivoting arm kickover tool:
 retrieving a valve with a positioning tool (Fig. 4.57)

This type of tool is indispensable in highly deviated wells and works as described below (Fig. 4.58).

The tool, which has a double articulation, is kept rigid in the axis of the setting string when running in (1). When pulling back up opposite the selected mandrel, a guide stop on the tool slides into a special sleeve placed in the upper part of the mandrel. This allows:

- the tool to be oriented properly with respect to the pocket (2)
- a spring to act and push the arm toward the side pocket (3).

In (4) the tool has latched onto the valve and in (5) the pivoting arm goes back into the upper narrow part of the mandrel and locks back into a rigid position after the valve has been retrieved. The assembly is pulled back up to the surface.

4.2.3 Surface equipment for a gas lifted well

4.2.3.1 Injection system

A. *Continuous gas lift* (Fig. 4.59)

The special gas lift equipment is usually confined to an adjustable choke on the gas inlet pipe upstream from the annulus valve. The adjustable choke supplies and controls the flow rate of injected gas.

B. *Intermittent gas lift* (Fig. 4.60)

Here the special equipment is a little more complicated since the aim is to get gas injections with a variable duration and frequency. Instead of an adjustable surface choke as in a continuous system, what is required is a device to control the periodicity and duration of injections. The device, called an intermitter, must also operate automatically.

4. ARTIFICIAL LIFT

Fig. 4.55 Kickover tool (*Source:* J. de Saint-Palais, J. Franc, *Cours de gas lift*, Éditions Technip, Paris, 1969).

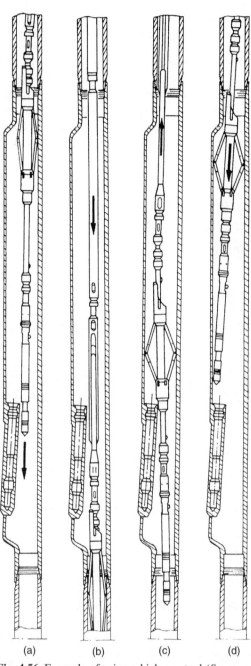

(a) (b) (c) (d)

Fig. 4.56 Example of using a kickover tool (*Source:* After a Flopetrol document).
a. Running in. b. Unlocking. c. Pulling up. d. Setting.

4. ARTIFICIAL LIFT

Fig. 4.57 Positioning tool (*Sources:* Teledyne Merla document, Composite Catalog 1986–1987).

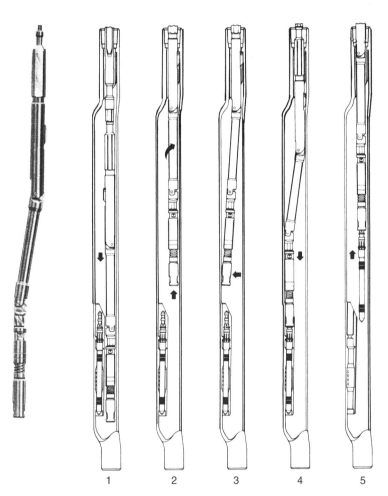

1   2   3   4   5

Fig. 4.58 Retrieval of a gas-lift valve (*Sources:* Teledyne Merla document, Composite Catalog 1986–1987).

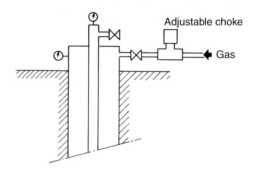

Fig. 4.59 Wellhead equipment for continuous gas lift (*Source:* ENSPM Formation Industrie).

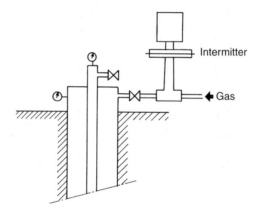

Fig. 4.60 Wellhead equipment for intermittent gas lift (*Source:* ENSPM Formation Industrie).

Principle of the Camco CT 26 intermitter

It usually consists of (Fig. 4.61):

- an automatic-operated valve (1)
- a pilot valve (2) for the automatic-operated valve
- a drum (3) driven by a clock to actuate the pilot valve.

Operation

The clock transmits its movement to a drum with small removable bars around it at regular intervals. The three-way control valve:

- lets gas in the valve servomotor (closing the valve), or
- opens the valve servomotor to the atmosphere (releasing gas, opening the valve).

4. ARTIFICIAL LIFT

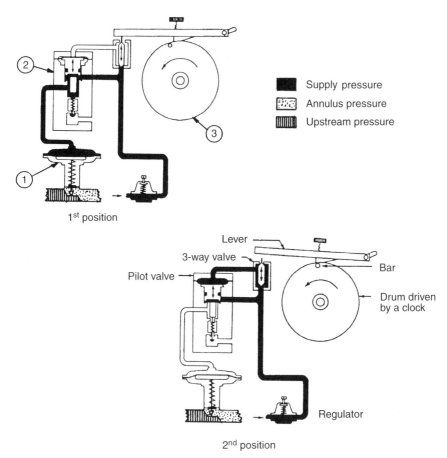

Fig. 4.61 Operating principle of the intermitter (*Source:* J. de Saint-Palais, J. Franc, *Cours de gas lift,* Éditions Technip, Paris, 1979).

The drum and the pilot valve are connected by a cam lever and this assembly is raised by the bars as each one passes against the cam. A screw device is used to vary the height of the cam. As a result:

- the automatic-operated valve remains closed as long as no bar is in contact with the cam
- it opens and gas is injected each time a bar passes against the cam.

The duration and periodicity are regulated in the following way:

- periodicity: bars are added to or removed from the drum in order to decrease or increase the distance between them
- duration: the cam height is varied, the higher the cam the longer the inclined part will remain on each of the bars and the longer the automatic-operated valve will remain open.

4.2.3.2 Measurements (Figs. 4.62 and 4.63).

If gas lift is to operate properly, especially during the start up phase, injection parameters, i.e. pressure and flow rate, must be monitored on the surface.

In addition to pressure and temperature indicators that normally equip the Christmas tree, the annulus injection line has a double tubing and casing pressure recorder. The injection flow rate can also be checked by installing a flowmeter. Measurements are usually made using a system with an orifice plate and a Barton type recorder.

4.3 CHOOSING AN ARTIFICIAL LIFT PROCESS

4.3.1 Economic criteria

The problem is to determine which artificial lift system will recover oil the fastest (discounting) in the largest amounts (recovering reserves) at the lowest cost (profitability). However, the overall cost of artificial lift (investment and operating costs) is not easy to assess.

As regards investment, though it is fairly easy to calculate the cost of specific artificial lift equipment (pump, pumping unit, compressor for gas lift, etc.), it is much more difficult to assess the proportion of extra cost due to the artificial lift process in the initial investment. For example the initial completion of the well is designed when possible taking into account the future artificial lift option that will be implemented later on. Likewise an offshore production platform is sized with due consideration to the extra equipment required for artificial lift.

As for operating expenses, although it is relatively easy to isolate direct operating and maintenance costs related to the artificial lift system during production, it is much harder to foresee them ahead of time. For example the energy expenditure specific to the artificial lift process presupposes that the energy efficiency of the process is known beforehand, and this is no easy job. Likewise maintenance and repair costs are usually based on statistical hypotheses (established in a given region of the world on a given field) which are not necessarily representative of the field under study.

Moreover, in the same way as for investment, it is not simple to allocate expenses incurred in treating commercial production (oil and gas) or non-commercial production (water, sand and sediments). Treatment can be influenced by the type of artificial lift chosen, as some processes for example give rise to emulsion formation or foaming, or facilitate injection of corrosion inhibitors or demulsifying agents downhole.

4. ARTIFICIAL LIFT

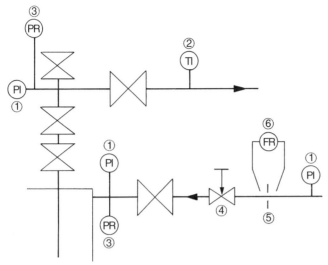

Fig. 4.62 Simplified diagram of the control and regulation equipment needed on the surface during well start up operations (*Source:* document from the "Artificial lift" seminar, ENSPM Formation Industrie).
1. Pressure indicator. 2. Temperature indicator. 3. Pressure recorder. 4. Manual gas flow adjustment choke. 5. Orifice plate. 6. Flow recorder.

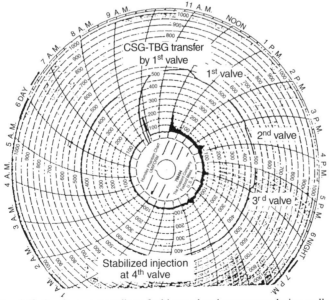

Fig. 4.63 Continuous recording of tubing and casing pressure during well start up (*Source:* document from the "artificial lift" seminar, ENSPM Formation Industrie).

4.3.2 Technical criteria

Choosing an artificial lift method requires a study of all possible processes beforehand without any preconceived notions. The aim is to determine which one is the most compatible with the requisite production specifications (wellhead pressure and flow rate mainly) and the constraints due to the reservoir and its environment. Three types of criteria are involved in the choice.

The first and simplest one is related to the energy source necessary for the process, its availability and access cost. From this standpoint processes with low energy efficiency will be at a disadvantage in regions where energy supply is expensive.

The second criterion is dual: pump pressure head (or target pressure) and liquid flow rate to be produced, with the product of the two representing the installed power that is needed. Note that the required pressure head is related to the bottomhole pressure and the depth of the well.

The other criteria take into account all the different operating constraints that stem from the following factors:

- The general environment. The state of the site, climatic conditions, the industrial environment, safety regulations and regulations concerning environmental protection are involved. As a result, the choice might favor sophisticated expertise and maintenance with better energy performance when qualified manpower is available for repairs and maintenance at an acceptable cost. In contrast, if there is associated gas in the field that can not be commercialized readily but provides a low-cost energy source, this will tip the balance toward processes with lower efficiency performance.
- The infrastructure on the surface and the immediate environment. The fact that wells are isolated or on the contrary grouped together in a cluster influences the choice of process. Ancillary surface equipment is usually substantial, so a number of factors need to be considered: available space and its cost, the number of equipment components, equipment maintenance and lifetime, risks of breakdowns, means of automation, standardization of parts, etc.
- The well architecture. Particularly the well profile, the available space, the number of levels to be produced separately and the well depth.
- The characteristics of the effluent that will be produced. Particularly temperature (related to the depth of the well), the water cut, presence of free gas, corrosive substances and sand, and viscosity.

4.3.3 Making a decision

In practice it is based on such qualitative and quantitative criteria that the choice of an artificial lift process is made. They are often difficult to assess and can vary with time. It is

4. ARTIFICIAL LIFT

not easy and care must be taken not to be misled by good or bad experience in the past with any given process. From one reservoir to another a large number of parameters can change and technology is progressing daily. Given the uncertainties (about how the field will perform, etc.), especially when the field needs artificial lift from the outset, a **temporary** system can be selected. Be sure it does not become **permanent** without having been reassessed!

4.3.4 Main advantages and drawbacks of artificial lift processes

The discussion is limited to the four most common artificial lift methods, i.e.:

- sucker rod pumping
- submerged centrifugal pumping
- plunger or jet hydraulic pumping
- continuous gas lift

excluding other processes which are currently more marginal such as:

- the Moyno pump
- the turbine pump
- intermittent gas lift.

4.3.4.1 Sucker rod pumping

A. Main advantages

- The most widespread technology, relatively simple and well known in the industry.
- Well suited to low and moderate flow rates (only if the pump is not too deep, however).
- The flow rate can be changed easily (rate, stroke), flexible operation.
- Compatible with very low bottomhole pressure.
- If there is a subsurface problem, it can be solved by a relatively lightweight servicing unit.
- Suited to isolated wells.
- Mechanical beam units are simple and durable, with low operating expenses as a result.
- Hydraulic units occupy little space and long stroke units are very useful for viscous and gassy crudes.

B. Main drawbacks

- Possible flow rate decreases severely with the depth required for the pump. Maximum performance is approximately 25 m^3/d (150 bpd) at 3000 m (10 000 ft) depth and 250 m^3/d (1500 bpd) at 750 m (2500 ft), depending on the maximum allowable stress on the sucker rods.

- Reduced volumetric efficiency in wells with high GORs.
- Beam pumping units take up too much space and are too heavy for offshore platforms.
- Initial investment cost is high for sophisticated large-capacity pumps, especially for hydraulic units.
- Major problem of sucker rod strength when there is a corrosive effluent.
- Ill suited to "crooked" well profiles.

4.3.4.2 Submerged centrifugal pumping

A. Main advantages

- High flow rates are possible at shallow or average depths.
- Well suited to production with a high water cut.
- Surface equipment takes up little space.
- Daily monitoring problems reduced to a minimum unless the pump breaks down.
- Good energy efficiency, advantageous if there is access to an existing cheap power network.

B. Main drawbacks

- Output capacity strongly influenced by depth.
- Limited in temperature and consequently in depth.
- Ill suited to low flow rates.
- Tubing must be pulled in the event of trouble (operating costs and costly downtime, especially offshore).
- Not usually recommended when the GOR is high or when wells are prone to gas breakthrough.
- Performs poorly in the presence of sand.
- Little flexibility except if variable-speed controler is used on the surface.

4.3.4.3 Hydraulic pumping

A. Main advantages

- Suited to great depths and deviated wells.
- Pump (depending on the installation) can be pumped up to the surface so that it is not necessary to pull the tubing (savings on handling operations and downtime).
- Working fluid can serve as a carrier fluid for injecting an additive (corrosion inhibitor, demulsifier, etc.).

And, for the plunger pump:

- The size and rate of the pump can easily be modified to adapt to well conditions.
- Viscous heavy crudes benefit from being mixed with a lighter working fluid.
- Production is possible with extremely low bottomhole pressures.

And, for the jet pump:

- High production flow rate is possible.
- No moving part inside the well.
- Only minor problems if sand or gas are present.

B. Main drawbacks

- Initial investment in surface equipment is quite high and its maintenance is fairly expensive.
- High pressure pump feed circuit (with consequent safety risks).
- Well testing causes problems, especially regarding assessment of produced fluids.
- Completion with multiple tubings may be required.

And, for the plunger pump:

- Rapid wear and tear on the pump if the fluid is corrosive or abrasive.
- Efficiency drastically lowered if free gas is present.

And, for the jet pump:

- Low efficiency, 25 to 30% (70% for plunger pumps).
- Need for bottomhole flowing pressure of over 3.5 MPa (500 psi), otherwise detrimental cavitation takes place in the flow nozzle.
- Is prone to form emulsions or foam.

4.3.4.4 Continuous gas lift

A. Main advantages

- Well suited to average or high flow rates.

    For example: from 300 to 600 m^3/d (1800 to 3600 bpd) in a 3 1/2",
    from 2500 to 4000 m^3/d (15 000 to 24 000 bpd) in a 7".

- Suited to wells with a good PI and relatively high bottomhole pressure.
- Well equipment is simple and gas-lift valves can be retrieved by wireline, operating conditions can be modified without having to pull the tubing.
- Initial investment can be low if a source of high pressure gas is available (if compressors need to be installed this is no longer true).
- No production problems when sand is present.
- An additive can be injected (corrosion inhibitor) at the same time as the gas.
- Suited to deviated wells.
- Well suited to starting up wells.

B. Main drawbacks

- Need for bottomhole pressure that is not too low, so sometimes the artificial lift method has to be changed at the end of the well's lifetime.
- The required injection gas volume may be excessive for wells with a high water cut.
- Need for high pressure gas, which can be costly and increases safety risks.
- Can not be applied if the casing is in bad shape.
- Gas processing facilities (dehydration, sweetening) can compound compression costs.
- Foaming problems may get worse.
- Surface infrastructure is particularly expensive if wells are scattered over a large area.
- Rather low efficiency, especially in a deep well.

Chapter 5

WELL SERVICING AND WORKOVER

The profitability of a well as an investment venture depends on how long it is on stream and on how much it produces. Its lifetime and output are naturally due to the reservoir's initial characteristics. However, they are also dependent on keeping the well maintained in good working order and adapting completion properly to the constantly varying conditions prevailing in the reservoir and around the wellbore.

The term well servicing covers all of the operations that can be performed on the well itself with either of two objectives:
- finding out how the status of the well itself or the reservoir is evolving
- maintaining or adapting the well to keep the best possible operating conditions.

By "the well itself" we mean the connection between the borehole and the pay zone, its immediate vicinity and everything that is located in the well up to and including the wellhead.

Processes such as artificial recovery, which deal with problems on the scale of the reservoir rather than the well, are not discussed here.

In addition, it should be remembered that the operations that can or must be done over the field's lifetime to keep the wells in good working order and profitable are largely influenced by the choice of the completion system done at the time of the initial completion.

5.1 MAIN TYPES OF OPERATIONS

The operations that may have to be carried out on a well are numerous and can be broken down into measurements, maintenance and workover. Measurements may involve the status of equipment, the quality of the pay zone-borehole connection or the status of the reservoir in the vicinity of the well. Maintenance and workover operations mainly affect equipment or the pay zone-borehole connection.

Maintenance is the relatively simple operations that can be done with the well still producing, i.e. under pressure, with lightweight means such as wireline units or a pump alone.

In contrast, workover operations entail using heavier means. They may sometimes be carried out with the well under pressure (using a coiled tubing or a snubbing unit for example), but usually require the well to be "killed" (i.e. placing a control fluid in the well whose hydrostatic pressure is greater than the reservoir pressure).

Servicing operations may be decided because of:
- operating considerations such as an abnormal drop in production, or a prematurely worn or obsolete equipment
- reservoir considerations such as knowing how the reservoir is evolving or how to best adapt to its behavior
- trouble that has cropped up when operations were carried out for the above-mentioned reasons, for example to retrieve a "fish" (any tool, piece of equipment or other item lost or accidentally stuck in the well).

5.1.1 Measurement operations

These may take place in a number of different locations.

5.1.1.1 At the wellhead

Here measurements mainly involve pressure and temperature (or even sample taking) at the wellhead and downstream from the choke. A variation in one or more of these parameters means that there has been a modification in production conditions (drop in reservoir pressure, plugging, variation in the percentage of water or gas, obstructions, etc.). The problem will be identified with the help of the other available information (measurements at the processing facilities, downhole, etc.), and measures will accordingly be taken if needed. In some instances wells are equipped with permanent bottomhole pressure sensors and so this parameter is directly available on the surface.

Pressure, or more exactly the lack of pressure, is also monitored at the top of the various annular spaces to check the integrity of casings, packers and production string (tubing). It should be remembered that during well clean up or when boosting the flow rate above the maximum rate attained before, the annular spaces (and all spaces where liquid may be trapped) have to be bled off.

For wells under sucker rod pumping, dynamometric measurements are used to monitor the stress on the rods and the pump's operating conditions. For wells under artificial lift by sucker rod pumping or by gas lift, echometers can be used to locate the liquid level in the annulus:
- for wells under rod pumping, pump submergence can be checked and bottomhole pressure can be assessed
- for wells under gas lift, it may be a help during start up or production (checking through which valve the injected gas passes).

Even though they may be considered as maintenance, we also mention surface and subsurface safety valve testing operations.

5.1.1.2 In the tubing

Measurements mainly consist of calibrations, to check that a wireline job is possible (running a pressure recorder, for example) or in connection with a problem of corrosion or deposits.

In a gas lifted well, temperature well logging can also be performed, with a recorder, to check that valves are working properly.

5.1.1.3 At the bottomhole

The measurements can be a check of the sediment top, with or without sampling, to make sure that a tool (recorder, etc.) can be run on wireline or to monitor sanding up of a well. They can simply be measurements of pressure or temperature (during production testing for example) at a specified depth, possibly made in conjunction with fluid (or bottomhole sediment) sampling for analysis.

Measurements may also be production logs: recording of flow rate, variations in the density of the effluent or the temperature, etc. all along the produced height. Logging is used qualitatively, and in simple cases quantitatively, to find out what each zone contributes in terms of flow and fluid type to overall production. For example a distinction can be made between a well that produces 50% water coming equally from all its perforations from one producing 50% water coming from only the lower perforations. Whereas in the second example the pay zone-borehole connection can be modified to lessen water production (by trying to blind up the lower perforations), in the first case modifying the connection will not solve the problem.

5.1.2 Maintenance operations

5.1.2.1 On the wellhead

In addition to routine operations such as adjusting the flow rate, opening or shutting in a well, we also include lubricating valves, replacing faulty parts downstream from the master valves, and periodic verification of surface and subsurface safety valve control systems (SSV = Surface Safety Valve, SSSV = SubSurface Safety Valve).

5.1.2.2 In the tubing and its equipment

These are operations connected with problems of deposits and/or corrosion such as cleaning the tubing by scraping, injecting a paraffin dispersant, a hydrate or corrosion inhibitor, etc.

Also included here is the injection right from the bottom of the well of chemicals such as demulsifiers, anti-foaming agents, etc. that make surface processing easier.

These operations may also be the exchange of equipment by wireline operations, such as a subsurface safety valve of the WireLine Retrievable type (WLR), or a gas-lift valve or the fishing up of any item left accidentally in the well during wireline or other jobs.

5.1.2.3 At the bottomhole and on the pay zone-borehole connection

These are the operations to be carried out:
- by wireline, such as cleaning out the bottomhole with a sand bailer, making further perforations, etc.
- by pumping from the surface, such as perforation acid wash, etc. (but, for an oil well, this means all the well effluent must be reinjected into the formation).

In actual fact the operations in this category often require more complex equipment and material and as a result will be dealt with in the following section.

5.1.3 Workover operations

Workover operations may be decided for a number of reasons.

5.1.3.1 Equipment failure

A. At the wellhead

What is involved here are mainly:
- leaks at the lower master valve, tubing hanger or tie-down screws
- a damaged Back Pressure Valve (BPV) seat
- problems at the SCSSV control line outlet: leak or failure.

B. In the subsurface safety valve system

The following cases may occur:
- a Tubing Retrievable (TR) SCSSV is faulty, or a wireline retrievable SCSSV is faulty and stuck
- the landing nipple of a wireline retrievable safety valve is leaky
- the control line is leaky or fails
- an annular safety system is faulty.

C. In the pipe

Whether in the casing or tubing, the problems are leaks (improper makeup, corrosion), and collapsed, burst or broken pipe. The tubing can also get partly or totally plugged up by deposits that can not be removed by conventional wireline jobs.

D. In the downhole equipment

Let us mention in particular the following:
- leaks on equipment that has sealing elements (packer, locator, slip joint, circulating sleeve, etc.)

- a packer that gets accidentally unseated
- wireline jobs that were not properly carried out: stuck gas-lift valve, wireline fish, etc.
- pumping problems (sucker rod or electric pumping): pump breakdown, broken rod, faulty cable, etc.
- miscellaneous faulty downhole equipment: permanent sensors, etc.

5.1.3.2 Modifications in production conditions

In order to get sufficient velocity to carry up the heavy phases (condensate or water in a gas well, water in an oil well) after a drop in flow rate, it may be advantageous to reduce the tubing diameter by changing the tubing or setting a concentric tubing. Otherwise, if the heavy phase is left to build up in the tubing, it may exert too much back pressure on the pay zone.

When the well's flowing capacity becomes insufficient an artificial lift process needs to be implemented or any existing process has to be modified. If in contrast the reservoir performs better than initially expected, a larger output can be contemplated. This may involve replacing the tubing by a tubing with a bigger diameter (if the casing size allows it) or modifying artificial lift (change in equipment or artificial lift process).

5.1.3.3 Restoration or modification of the pay zone-borehole connection

This type of operation may be warranted in order to:

- stimulate (acid job or fracturing) a zone that is producing less than expected
- implement or restore sand control
- bring a new zone on stream
- try to limit unwanted fluid inflow (water and/or gas for an oil reservoir, water for a gas reservoir) by remedial cementing, by isolating perforations or abandoning a zone
- restore cementing to avoid communication between formation layers, etc.

5.1.3.4 Change in the purpose of the well

When the conditions in a field have evolved, particularly when water/oil, oil/gas or water/gas contacts (or transition zones) have progressed, a production well may be turned into an injection or observation well (or vice versa) while the same formation is still being produced. However, this does not necessarily entail a workover operation.

After considering field evolution, it may also be decided to shut off a producing zone that is more efficiently drained elsewhere, while opening a formation that had previously been ignored. Depending on the well configuration this may be done simply by operations in the existing well, but it may also require deepening the well or sidetracking it.

Finally, the well may have to be abandoned temporarily or permanently.

5. WELL SERVICING AND WORKOVER

5.1.3.5 Fishing

When measurement, maintenance or workover operations are carried out, "fish" may accidentally be left in the well. The problem is then to attempt to retrieve it.

5.2 LIGHT OPERATIONS ON LIVE WELLS

The basic method of servicing a live well is by wireline. Other methods (such as pumping, etc.) are sometimes used.

5.2.1 Wireline work

5.2.1.1 Principle and area of application

Wireline is a technique used to operate in a producing or injecting well by means of a steel cable, usually a slick line, to enter, run, set and retrieve the tools and measurement instruments needed for rational production.

The advantages of this technique are important:

- work can be done inside the tubing without killing the well, by means of a lubricator connected to the wellhead, operations can be carried out under pressure and even without stopping production
- operations are performed quickly due to the use of lightweight, highly mobile equipment and run by two or three specialized operators
- money is saved because of the two points here above:
  - production is hardly stopped or not stopped at all
  - the pay zone is not damaged during operation (it is not necessary to kill the well)
  - simple material and limited human resources are used, consequently it can be readily implemented at relatively low cost.

There are, however, a number of drawbacks and limitations:

- the work requires highly qualified personnel
- it is risky to work in wells that are highly deviated, produce a lot of sand or have a viscous effluent
- operations are impossible when there are hard deposits
- the possibilities afforded by the wire are limited (it can only work in tension and at very moderate loads, no rotation or circulation is possible).

Wireline jobs can be classified into three different types:

- checking and cleaning the tubing or the bottomhole (checking inside diameter, corrosion, clogging, sediment top, etc.)

5. WELL SERVICING AND WORKOVER

- carrying out measurements (bottomhole temperature and pressure recordings, sampling, locating interfaces, production logging, etc.)
- running or retrieving tools and operating in the well (setting and pulling subsurface safety valves, bottomhole chokes, plugs, gas-lift valves, etc.; shifting circulating sleeves; fishing, perforating).

Some tools run into the well need an electric cable, in this instance the term is electric wireline.

The necessary surface and downhole equipment is covered in the following sections.

5.2.1.2 Surface equipment

Whatever the operation it will always be necessary to bring the following standard surface equipment to the well site (Fig. 5.1):

- The winch with a drum, the basic component where the wire is spooled up, which is driven by a motor or an engine; a depth indicator to monitor the wire going in and coming up.
- The lubricator, which is pipe of variable length equipped with a stuffing box sealing around the wireline on one end and a "quick union" on the other, allows tools to enter or be removed from well under pressure:
  - the lubricator is placed above the swab valve vertically in the well axis by means of a gin pole fixed to the wellhead by chains and collars beforehand,
  - the lubricator is equipped with a valve to bleed it off at the end of the operation,
  - between the base of the lubricator and the wellhead the following are located:
  a) a BOP type safety valve closing around the wireline
  b) sometimes a tool trap to hold back the tools in the event the wire should fail at the wireline socket, due to a faulty maneuver at the end of pulling out in the lubricator.
- The tensiometer, or weight indicator, that indicates the tension on the wire at the winch at all times.
- Miscellaneous equipment and material used in conjunction with this surface equipment such as: hoisting block, hay pulley, clamp, bleed-off hoses, etc.

A. The winch and its motor or engine (Fig. 5.2)

The basic component of the wireline winch is a steel drum that is large enough to hold a good length of wire. The winch is equipped with a depth indicator that constantly shows the depth of the tool hanging on the wireline with respect to a reference or zero point such as the upper tubing-head spool flange or the master valve. The principle of the depth indicator is usually as follows. At the exit of the winch, the wire drives a wheel of known circumference by friction. A counter-totalizer transforms the number of wheel rotations representing the unspooled length into feet or meters (attention: a wheel corresponds to a specific wire diameter, additionally the groove of the wheel gets deeper with use and this throws the measurement off).

5. WELL SERVICING AND WORKOVER

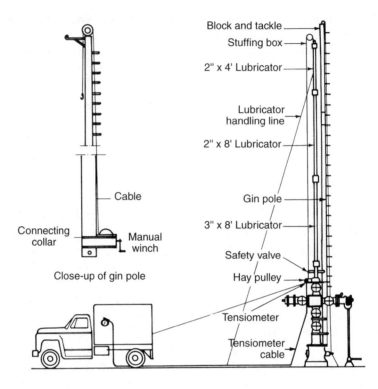

Fig. 5.1 Surface equipment (*Source:* G. Gaillot, *L'équipement de surface en travail au câble,* Éditions Technip, Paris, 1984).

Fig. 5.2 Hydraulic winch and its engine (*Source:* G. Gaillot, *L'équipement de surface en travail au câble,* Éditions Technip, Paris, 1984).

5. WELL SERVICING AND WORKOVER

The winch is driven by a diesel engine (or a gasoline engine for low rated output) or sometimes by an electric motor. The engine's rated output (and the wire diameter) depends on the working depth and on whether or not jarring will be used. Table 5.1 gives an idea of some orders of magnitude.

Table 5.1.

Winch horsepower (HP)	Recommended maximum depth			
	without jarring		with jarring	
	(m)	(ft)	(m)	(ft)
9	2 000	6 700	500 (by hand)	1 700
14	3 000	10 000	2 000	6 700
22	5 000	16 700	2 500	8 300
48	5 000	16 700	5 000	16 700

Between the engine and the drum there is a transmission which is generally mechanical or hydraulic. The advantage of a mechanical transmission is that it is simple and sturdy. Meanwhile the major drawbacks are that its speed range is fixed once and for all and there is no safety system in the event the wireline string gets caught when pulling out. Hydraulic transmission, increasingly preferred to the mechanical version, uses a moving fluid to drive the winch in the required direction via a hydraulic motor. It allows:
- regular, steady drive
- flexible, progressive clutching
- continuous braking with no overheating, since the oil cools down all throughout the circuit
- a wide range of speeds
- a tensile stress safety system on the wire by means of a relief valve that limits the pressure in the hydraulic circuit.

Other features, such as a brake on the drum or in some instances a gear box, are also included in the winch system. Table 5.2 gives an idea of the recommended speeds for a number of operations.

Table 5.2.

Operations	Running in	Pulling out
Amerada recorder	1 m/s (3 ft/s)	1 m/s (3 ft/s)
Sampler	1 m/s (3 ft/s)	maximum
Tubing/bottomhole check	2 m/s (6 ft/s)	2 m/s (6 ft/s)
Setting mandrels	Depending on well	Depending on well
Paraffin removal	Depending on well	Depending on well
Caliper	Unimportant	20 to 22 m/min (70 ft/min)

5. WELL SERVICING AND WORKOVER

There are also winches with two drums or double drums. One spool is equipped with slick line and the other with braided line which is used mainly for operations requiring considerable tensional strength. A simple switching system is used to select one or the other of the drums to be driven by the same engine.

B. The cable

The wireline cable is usually a slick steel wire of the "piano string" type, drawn in one piece without any welding or brazing in accordance with API SPEC 9 A standards (API: American Petroleum Institute; SPEC: specification). The most common diameters are: 0.066", 0.072", 0.082", 0.092" and 0.105" (i.e. from 1.7 to 2.7 mm). There is also a 3/16" braided or stranded line for jobs requiring greater tensional strength (swabbing, etc.). This cable needs special devices (rope socket, safety devices for the lubricator). Wireline cable is delivered in reels ranging from 10 000 to 25 000 feet (3000 to 7500 m). The weight is normally expressed in weight per length, i.e. mainly in pounds per 1000 feet or in grams per meter.

Three types of wire are commonly used:

- ordinary steel (improved plow steel or IPS) with good mechanical properties
- galvanized steel which is brittle in a chlorine environment
- stainless steel (among which the Uranus 50 wire) with good H_2S resistance but rapid strain hardening.

For example, the 0.072" (1.83 mm) ordinary steel wire has a weight of 17 g/m (11 lb/10^3 ft) and a breaking load of 4.4 kN (985 lb). For a 0.092" (2.3 mm) wire, the figures are 33 g/m (22 lb/10^3 ft) and 7 kN (1590 lb) respectively. The breaking load of stainless wire amounts to only about two-thirds of this.

Since the wire is subjected not only to tensile stress but also to bending stress (bending stress beyond its yield strength at the hay pulleys) during an operation, it is recommended to:

- restrict tension on the wire to 50% of its minimum breaking load (and even less in very hot wells)
- cut a piece of the wireline after a long jarring operation (four hours) if more work is to be done at the same depth (so that it is not always the same stretch of wire that fatigues on the pulley system)
- scrap the wire if it no longer meets API test specifications or if it shows evidence of having lost elasticity (the cable at rest no longer tends to curl up in spirals, seems soft when attached to the socket, etc.) or if it has been used for over 50 hours of mechanical jarring.

C. The weight indicator or tensiometer

These devices show at all times how much tension is being exerted on the wire, thereby avoiding overloading which causes fatigue and eventually wire failure. Consequently, they are of capital importance.

5. WELL SERVICING AND WORKOVER

The most conventional tensiometers are:
- hydraulic: tension on the wire compresses a liquid via a deformable membrane and transmits the pressure variation to a pressure gage that is directly graduated in pounds, kilogram-force or newtons (Martin Decker)
- electric: tension on the wire causes variations in the electrical resistance of a potentiometer in a pressurized hydraulic chamber (Bowen Itco) or that of a strain gage placed on the pulley of the depth indicator (Flopetrol).

Note that for the first two devices (installed on the hay pulley connection at the foot of the mast) measurement depends on the angle made by the wire on the pulley, they are calibrated for an angle of 90°.

D. The lubricator

A steel lubricator is placed on top of the Christmas tree to allow tools to be run into a pressurized well. It serves as an intermediary between the well and the surface environment. The lower part of the lubricator has a quick union on the end and sometimes a tool trap before it. The quick union screws onto the BOP placed above the swab valve. The stuffing box is usually included under the term lubricator, it is through the stuffing box that the wireline enters the well.

To allow easy transportation and handling the lubricator is made up of several sections that are connected together by quick unions with a seal provided by O rings. Common sections are 8 feet long (2.40 m) and the number of sections depends on the length of string that is going to be run into the well. This can range from one section for simply running in an Amerada clock to four or five for fishing out strings left downhole. The most widespread diameters go from 2" to 4" (50.8 to 101.6 mm) and the series from 3000 to 10 000 psi (21 to 70 MPa).

It is at the base of the lowermost section that a 1/2" bleed off valve is placed which is used to bleed off the lubricator after the swab valve or the BOP has been closed. Before lifting the lubricator up vertically, the wireline is run through the stuffing box and then connected to the rope socket. All that remains to be done is to screw the tool string onto the rope socket before connecting the lubricator to the well.

E. The stuffing box

This device is placed on the uppermost end of the lubricator and allows the cable to pass through into the well while providing a seal at the same time.

It consists of three assemblies (Fig. 5.3):
- a lower quick union
- a body with packings that can be compressed against the cable by means of the packing nut, and a mobile safety plunger with a sealing element that comes to rest against the main packing lower gland and seals against it if the wire should fail

- a sheave and its holder that can rotate on the lubricator's axis so as to line up with the hay pulley.

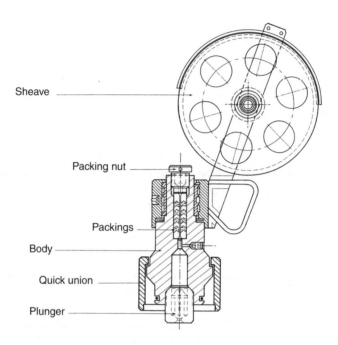

Fig. 5.3 Stuffing box (*Source:* G. Gaillot, *L'équipement de surface en travail au câble,* Éditions Technip, Paris, 1984).

F. The tool trap

This is a safety system located at the base of the lubricator. It consists simply of a tilting device placed above the BOP that keeps the wireline string from falling into the well should the wire break due to the impact of the rope socket against the stuffing box. This type of accident can be caused by a wrong move or by a faulty depth measurement indication.

G. The wireline valve (BOP) (Fig. 5.4)

Located between the lubricator and the top of the wellhead, the wireline valve is used to quickly isolate the first from the second by closing around the wireline. It is often necessary to have a wireline valve for safety reasons or for special operations such as wire fishing. In contrast with the stuffing box, this safety valve, commonly called BOP (blow out preventer) is designed to be used under static conditions (wire at rest).

Models differ from one manufacturer to another, but the basic principle consists in rams with rubber sealing elements that press against one another tightly around the wire. The rams are actuated either hydraulically or with a handwheel.

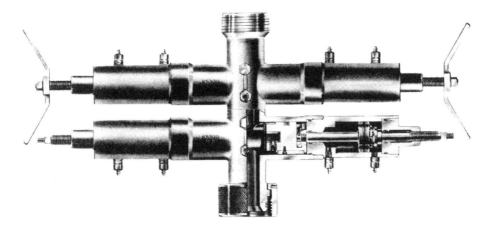

Fig. 5.4 Double BOP (*Source:* G. Gaillot, *L'équipement de surface en travail au câble,* Éditions Technip, Paris, 1984).

H. Special equipment for stranded line

Whether the cable is conventional strand (to give better tensile strength for example) or electrical strand (for production logging), special devices are requested, for example:

- flow tube control heads, or grease/oil injection control heads
- special dual BOP (holding the pressure in the upward direction, for the upper one, and in the downward direction for the lower one, with grease injection between them).

5.2.1.3 The wireline tool string

The tool string is all the equipment placed above the specialized tools (e.g. for checking and maintenance, setting and retrieving, fishing, etc.) and is used to run them in hole. It is connected to the wire by a special socket.

Even though the string's makeup depends on the job that is going to be done and the well conditions, the tool string usually consists of:

- a rope socket
- stems
- a spang jar
- a knuckle joint.

Specific equipment can be added such as a quick-lock coupling, a hydraulic jar, a kick-over tool, etc.

All of the components in the wireline tool string have a standard fishing neck. They are connected to each other by sucker rod type threads that must never be greased or oiled and are blocked energetically by hand with special tongs. The diameter of the wireline string is chosen according to the inside diameter of the tubing and its accessories. For example, the nominal 1 1/2" diameter (38 mm) is well adapted to 2 7/8" and 3 1/2" tubings in particular.

A. *The rope socket* (Fig. 5.5)

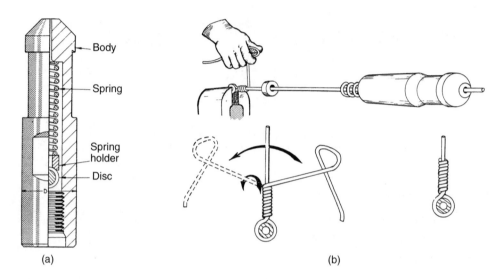

Fig. 5.5 a. Rope socket. b. Tying the knot (*Source:* IFP, Report No. 10404, 1964).

This socket is threaded at its lower end and connected to the wireline by means of a special knot tied inside (the wire is rolled up after being bent around a rolling disc). A spring placed in a machined cavity in the center serves as a shock absorber when jarring.

B. *Wireline stems or sinker bars* (Fig. 5.6)

These are heavy bars that allow the string to be run into the well despite the wellhead pressure and friction. They also serve as percussion weights to accentuate the jarring force. The stems' length and diameter are chosen according to the job that is going to be performed. The length is, however, limited by the available length in the lubricator and by the strength of the wire. They are delivered in three lengths: 2, 3 or 5 feet (0.61, 0.91 or 1.52 m).

5. WELL SERVICING AND WORKOVER

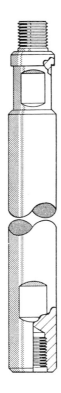

Fig. 5.6 Stem (*Source:* IFP, Report No. 10404, 1964).

C. Jars (Fig. 5.7a and 5.7b)

Jars make it possible to hammer with the stems at the end of the stroke in order to shear a pin in a tool (running, pulling, etc.) or for use in a maintenance or fishing operation (scraping, etc.). No jar is used when measurement instruments such as pressure recorders, etc. are run. Their stroke length is 20 or 30 inches (0.51 or 0.76 m).

Mechanical jars (spang jars or tubular jars) can jar upward and downward by sudden pulling or slacking off the wireline to accelerate the stems. Tubular jars are sturdier but give a less straightforward impact.

Hydraulic jars only allow upward jarring. First the wireline is tensioned: in the jar a plunger moves slowly in an adjusted sleeve before getting to an enlarged section which causes the movement to accelerate. This gives a powerful impact despite the friction in the well. Since the elasticity of the wireline is used in hydraulic jarring, these jars can not be employed at shallow depths (less than 500 m, or 1600 ft). They are always used in conjunction with a mechanical jar (generally located above it) to allow downward jarring as well.

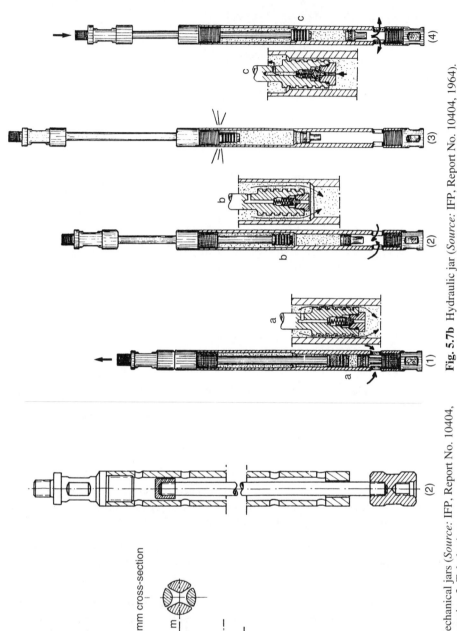

Fig. 5.7b Hydraulic jar (*Source*: IFP, Report No. 10404, 1964). 1. Oil flow restricted. Tensioning the wireline. 2. End of flow restriction. Stem acceleration. 3. Impact. 4. Closing the jar.

Fig. 5.7a Mechanical jars (*Source*: IFP, Report No. 10404, 1964). 1. Spang jar. 2. Tubular jar.

D. The knuckle joint (Fig. 5.8)

Because of its joint — though the angle is limited — the knuckle joint can provide flexibility to the tool string (centering tools, deviated wells, etc.). In addition, it keeps the wire from twisting around due to the tool string dragging against the tubing walls when it is run in and pulled out. If forceful jarring is planned, it is better not to use a knuckle joint. Note that there is a knuckle jar with a short stroke (about 12 cm, or 5") that can serve as a joint only in the open position.

E. Miscellaneous components

There are other components that can be added to the wireline tool string as described below:
- A quick-lock coupling consists of two parts that are readily connected by a quarter turn system locked by a spring (Fig. 5.9). The upper part is screwed onto the tool string and the lower part to the tool to be run into the well. It is better not to use this type of coupling if forceful jarring is planned.
- Kickover tools are used to set or retrieve valves in sidepocket mandrels (particularly in gas lift).
- Crossover subs.

5.2.1.4 Wireline tools

There are a large number of tools, broken down into four categories below:
- checking and maintenance tools
- running and pulling tools
- lock mandrels, downhole tools and other particular tools
- fishing tools.

A. Checking and maintenance tools

These tools are screwed directly at the bottom of the wireline tool string and are designed to check and clean the inside of the tubing and the bottom of the well. Let us mention a few in particular:
- Gage cutters are run in prior to any other wireline operation to check that the way is clear. The shape of the cutter allows some types of deposits to be scraped clean (Fig. 5.10).
- Swaging tools are designed to straighten tubing walls that have been slightly deformed locally. They should be used with caution because there is a risk of getting them stuck. (Fig. 5.11).
- Scratchers are used to "sweep" the tubing. In the same way as when gage cutters are used to clean the tubing, the debris has to be removed by having the well flow while scratching is being done to limit the risk of sticking (Fig. 5.12).

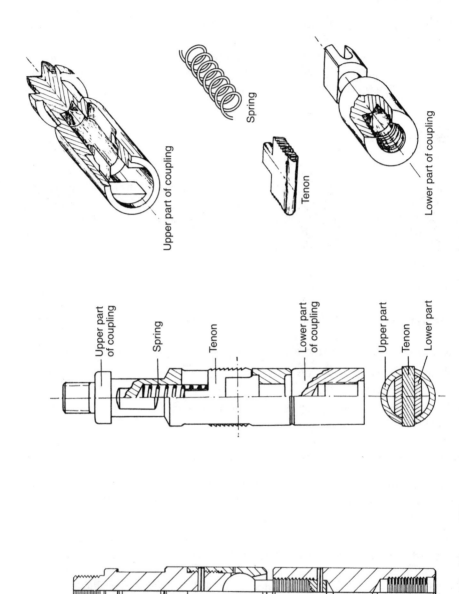

Fig. 5.9 Quick-lock coupling (*Source*: IFP, Report No. 10404, 1964).

Fig. 5.8 Knuckle joint (*Source*: IFP, Report No. 10404, 1964).

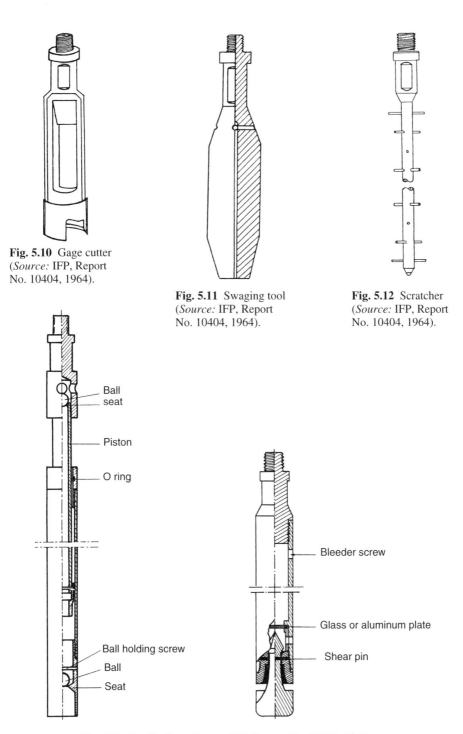

Fig. 5.10 Gage cutter (*Source:* IFP, Report No. 10404, 1964).

Fig. 5.11 Swaging tool (*Source:* IFP, Report No. 10404, 1964).

Fig. 5.12 Scratcher (*Source:* IFP, Report No. 10404, 1964).

Fig. 5.13 Sand bailers (*Source:* IFP, Report No. 10404, 1964).
a. Mechanical. b. Hydraulic.

- Calipers are equipped with feelers and record the variations in inside diameter (corrosion, deposits, etc.). Depending on the model they record the maximum defect found on a cross-section or defects along several generating lines.
- Sand bailers are used to take samples of sediment fill at the bottom of the well or even to clean the bottom of the well or the head of a tool that needs to be retrieved. They are mechanical (similar to a valve pump) or hydraulic (atmospheric pressure chamber protected by a device that is sheared by jarring once at the bottomhole). They are equipped with a check valve (Figs. 5.13a and 5.13b).

B. Running and pulling tools

These are specialized tools for installing and retrieving downhole tools. Downhole tools are attached to the tool string by means of the standardized landing and fishing heads located on the upper part of each tool. Running and pulling tools work by shearing one or more pins.

A distinction can be made between:

- retainer pins (tangential or transverse pins) which attach the downhole tool directly to the running tool
- shear pins, internal pins in the running and pulling tool mechanism, which release gripping dogs under normal operating conditions or as a safety precaution.

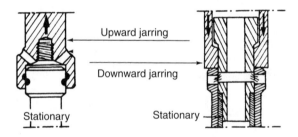

Fig. 5.14 Examples of shearing (*Source:* IFP, Report No. 10404, 1964).

Running and pulling tools can be classified into three categories:

- Running tools that run in and set downhole tools in the well, holding onto them by the retainer pins or by gripping dogs. They allow the downhole tools to be landed or locked in place, then to get free of them by jarring (upward, downward or both depending on the way the running tool and the downhole tool work) (Fig. 5.15).
- Pulling tools that grip and retrieve downhole tools operating in the well up to the surface. Elastic gripping dogs are used to latch onto the fishing head on the downhole tool, which can then be unseated and pulled up by jarring in a predetermined direction. Jarring in the opposite direction is used to break free of the downhole tool after shearing a pin if the tool will not unseat (Fig. 5.16).

5. WELL SERVICING AND WORKOVER

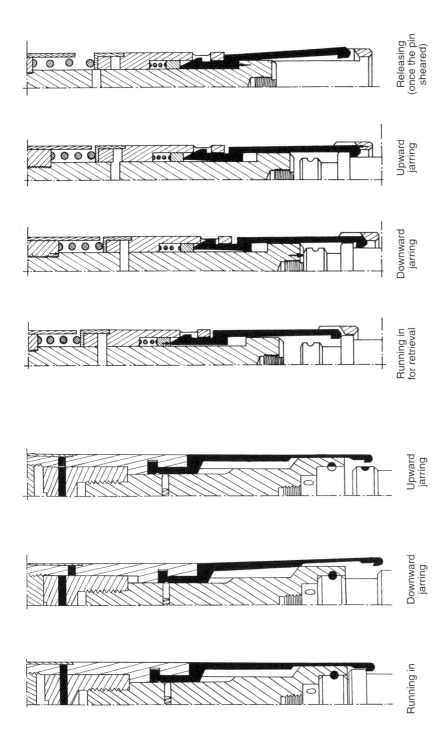

Fig. 5.16 Example of a pulling tool (*Source*: IFP, Report No. 10404, 1964).

Fig. 5.15 Example of a running tool (*Source*: IFP, Report No. 10404, 1964).

- Combination tools that fulfill the two functions (with a different procedure and sometimes with different equipment). These tools are usually adapted to one single type of downhole tool.

Before using running and pulling tools it is advisable to:

- know the characteristics of the downhole tool that is going to be run in or pulled out
- check that the tool works and that it matches the downhole tool prior to running it in;
- insert the pins that are suited for the work that is going to be done
- be sure that the tool required for pulling the downhole tool is available before running it in.

C. Lock mandrels, downhole tools and other particular tools

We mention here only some tools that have for the most part been referred to in previous chapters:

- Lock mandrels that carry the tools screwed on under them. They are landed and locked into landing nipples integrated in the tubing or sometimes directly on the tubing walls.
- Downhole tools such as plugs, equalizing subs, bottomhole chokes, subsurface safety valves and tubing pressure test tools. Equalizing subs are inserted between the mandrel and the plug as such and allow pressure differential to be equalized across the plug assembly (mandrel, equalizing sub, plug) before it is unseated and retrieved.
- Tools to operate circulating devices.
- Kickover tools to work in sidepocket mandrels.
- Special instrument hangers that can be utilized without any jarring for recorders.
- Swabbing tools to start up or kick off the well.
- Perforators which are used to make the tubing and the annulus communicate (usually before workover in order to neutralize the well or gas lift a well that is not appropriately equipped). There are mechanical perforators (a punch is beaten through the tubing by jarring) and gun perforators (an explosive charge is fired by jarring). They allow to install a calibrated port.

D. Fishing tools

Despite all the careful precautions during operations, tools and tool strings sometimes happen to get stuck and the wireline breaks. Before resorting to more complex means of action, a number of wireline tools may sometimes solve the problem by wireline techniques.

Here again there are a large number of tools, in particular those that might be used are as follows:

- Wireline cutters, designed to cut the wire flush with the rope socket when the wireline string is stuck. They are dropped from the surface and the impact on the rope socket is used to shear the wire. When the wire has knotted up into a "nest" (see next tool) and the wireline cutters can not get down to the rope socket, make sure to use a special wire-

cut the wire at the top of the nest (first send the conventional wireline cutter itself and then a go-devil, an instrument dropped inside the tubing that bumps against the wireline cutter).
- Wireline finders, used to find the upper end of a broken wireline and tamp it down (by light jarring) to make it into a nest so that it is easier to fish out with a wireline grab (see following tool). Finders are bell shaped with a diameter as close as possible to the tubing diameter (to keep the wire from sliding by to one side). They have holes in them with a diameter smaller than the wire diameter (to let the fluid through). Be careful not to run them lower than the end of the cable. (*Note:* when a wireline fails it falls only a little way down in the tubing, approximately 1 m per 1000 m, or 1 ft per 1000 ft in a 2 3/8" tubing).
- Wireline grabs, which serve to catch the wire and bring it up to the surface. They consist of two or three branches with teeth and the diameter should correspond to the inside diameter of the tubing (run in only after making an attempt with a finder) (Fig. 5.17).
- Impression blocks, designed to identify the shape and condition of the head of the fish. They are bell shaped and filled with lead. (Fig. 5.18).
- Overshots, which allow certain types of broken equipment to be fished. They feature a bowl equipped with a grapple (basket, spiral) that latches onto the head of the fish from the outside. (Fig. 5.19).
- Magnets, used to fish out small pieces of steel. They are protected by a sliding skirt made of antimagnetic metal when run in hole.

Fig. 5.17 Wireline grab (*Source:* Flopetrol).

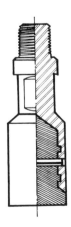

Fig. 5.18 Impression block (*Source:* Flopetrol).

Fig. 5.19 Overshot (*Source:* Flopetrol).

5.2.2 Pumping

A pump can be connected to the wellhead in order to inject a treatment fluid into the tubing or the vicinity of the wellbore (corrosion inhibitors, acid for washing perforations, etc.).

In fact this practice, which may seem simple at first, is not usually well suited to oil wells. The effluent that was initially in the tubing must be forced back into the pay zone and this is not necessarily easy (no injectivity) or may damage the pay zone. Otherwise a circulating device at the bottom of the well would have to be opened beforehand. However, in this case:

- Depending on the circulating device, the type of effluent, the temperature and on how long the equipment has been installed, the device may no longer be tight when it is closed again.
- With direct circulation, the effluent is pumped into the annulus and this will pollute the annular fluid.
- With reverse circulation, all of the annular fluid must first be circulated out and then replaced afterwards. In addition, some of the treatment fluid may be trapped by gravity in the annulus between the packer and the circulating device and this can cause problems, if it is an acid for example.

However, this practice can be advantageous for gas wells, which usually have fewer injectivity problems, or for wells where the treatment fluid can move downhole by gravity by migrating through the gas.

5.3 HEAVY OPERATIONS ON LIVE WELLS

We have seen that operations that can be carried out by wireline are limited in particular by the fact that:

- it is impossible to circulate
- it is impossible to put weight on the downhole tool or rotate
- the wireline tensile strength is low.

Other techniques allow work to be done on pressurized wells in the same way as wireline does and offset some of these drawbacks to a certain extent. However, they involve larger scale operations in terms of manpower and equipment and they can not be mobilized as readily. Two of these techniques are coiled tubing and snubbing.

5.3.1 Coiled tubing

5.3.1.1 Principle and area of application

The coiled tubing unit (Fig. 5.20) consists of continuous metal pipe with a diameter of 3/4" to 1 1/2" (19 to 38 mm) or even more, coiled up on a spool (or reel). It can be run in and pulled out of a pressurized well. The pipe, usually with its end equipped with a check valve at least, is maneuvered via an injector head through a safety system. The operation requires a specialized team of at least three people.

Fig. 5.20 Coiled tubing unit (*Source:* Otis document, Composite Catalog 1990–1991).

Coiled tubing units can carry out a number of operations on pressurized wells rapidly (lightweight rig, no pipe lengths to be made up, etc.) and are first and foremost a means of circulating in the well. They are therefore used in particular to:

- make the hydrostatic column lighter prior to perforating (underbalanced perforating after equipment installation)
- start up or kick off a flowing well (after a stimulation job for example) by circulating a "light" fluid (i.e. exerting a hydrostatic pressure that is lower than the reservoir pressure) or by injecting nitrogen
- implement temporary gas lift (during testing while drilling, when waiting for workover, etc.)
- reduce, hence optimize the flow path through the tubing (in a well with heavy phase segregation problems after the flow rate has dropped)
- clean out the tubing (sand, salt, paraffins, hydrates, etc.) by circulating an appropriate fluid (water, brine, hot oil, alcohol, etc.)
- clean out the bottom of the well by circulating (sand fill, etc.)
- spot acid, solvents, etc. opposite the zone(s) that need to be treated
- inject a killing fluid by circulating (prior to workover, etc.).

Depending, for some of them, on the coiled tubing diameter, they are also used to:
- spot cement plug
- clean out hard fill and scale, with a jetting tool or a drill bit associated with an hydraulic motor
- do lightweight drilling out (cement plug, etc.)
- perform some fishing job (with an overshot, etc.)
- set a "temporary" concentric tubing to inject inhibitor, for gas lift, etc.
- perform some drilling with an hydraulic motor (deepening, making an horizontal drain in a conventional well, etc.).

Let us mention the special case of horizontal wells where coiled tubing can be used to get tools into a horizontal drain hole, in particular for well logging (in which case the electric cable will have been placed in the coiled tubing before it is spooled up on the reel).

5.3.1.2 Coiled tubing equipment (Fig. 5.21)

Besides the tubing itself the equipment in a coiled tubing unit consists principally of the following:
- a reel
- an injector head
- a safety assembly
- ancillary surface equipment: a cab, a power pack, a crane, etc
- downhole accessories.

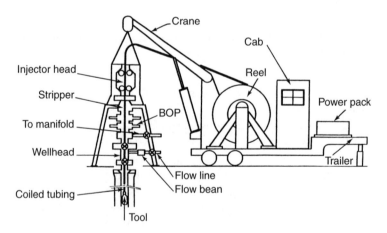

Fig. 5.21 Coiled tubing installation (*Source: Forage,* No. 120, 1988).

A. *The tubing*

The tubing consists of a strip of mild steel rolled cylindrically and welded longitudinally. Unit lengths of several hundreds of meters (about a thousand feet) are butted together by radial welding to make up coils up to 6000 m (20 000 ft) long. The outside diameter is 3/4", 1 1/4" or 1 1/2" (19, 25, 32 or 38 mm) and even more.

B. The reel

The reel, with a diameter of about 2.5 m (8 ft), can contain approximately 6000 m (20 000 ft) of 1" tubing. It is driven by a hydraulic motor. The tubing coiled up on the reel is connected by a reel swivel to the rotation axis. This allows the connection with a pumping system and enables pumping to continue while the reel is coiling or uncoiling the tubing in the well. There is a system to sprinkle corrosion inhibitor onto the tubing when it is pulled out of the well.

C. The injector head (Fig. 5.22)

The injector head uses a friction drive system actuated hydraulically (a series of half slips contoured according to the tubing diameter and mounted on two chains).

Depending on the wellhead pressure, it must be able to:

- push the tubing into the well as long as the tubing weight is not greater than the force exerted by the pressure in the well on the total tubing cross-section
- then to support it.

It is also equipped with a guide for the tubing (or gooseneck), a straightener, a depthometer and a weight sensor.

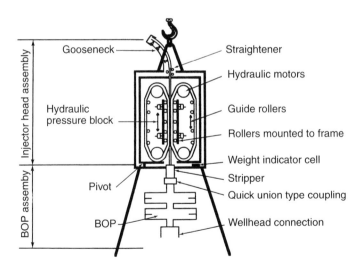

Fig. 5.22 Injector head (*Source: Forage*, No. 120, 1988).

D. The safety assembly (Fig. 5.23)

The assembly consists of a stripper (components with sealing elements) that provides a seal in the dynamic phases (tubing being run in or pulled out) and a stack of ram-type BOPs that fulfill the safety function in the static phases.

The stripper (Fig. 5.24) is located above the BOPs. The sealing element is actuated hydraulically at a pressure that depends on the wellhead pressure. The element is in two parts and can be changed during tripping after having closed the relevant BOP.

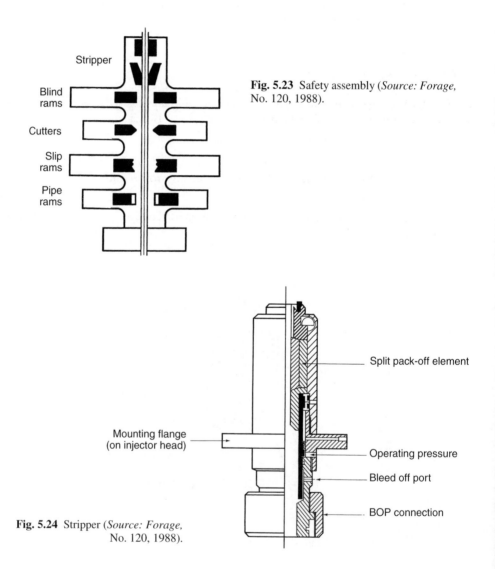

Fig. 5.23 Safety assembly (*Source: Forage*, No. 120, 1988).

Fig. 5.24 Stripper (*Source: Forage*, No. 120, 1988).

The BOP stack is made up of four stages, from bottom to top:

- pipe rams (that close and seal on the tubing)
- slip rams (that hold the tubing)
- cutters (that cut the tubing)
- blind rams (that give a total closure over a cut tubing).

Furthermore, there are also valves to equalize the pressure on either side of the rams and to allow pumping through the tubing if it has been cut off, thereby placing the well in safety conditions.

E. Ancillary surface equipment

Other equipment is required such as:

- A hydraulic crane with a collapsible boom or a hydraulically actuated mast to install and retrieve the safety assembly and the injector head.
- A power pack consisting of a diesel engine driving a hydraulic pump that provides the energy required by all the equipment.
- An adjustable height control cab placed behind the reel in order to have a direct view of the reel, the injector head and the safety assembly. All of the controls, gages and warning lights needed to operate, monitor and keep the unit safe as a whole are located there.
- If need be, a nitrogen unit, i.e. in particular low temperature storage tanks for liquid nitrogen, a pumping unit with a cryogenic booster pump and a heat phase converter.

F. Downhole accessories

In all operations where the coiled tubing is exposed to the well's pressure and fluids it is recommended to use check valves mounted on the lower end of the tubing (at least two for safety). This helps limit the risks in the event of a leak at the reel swivel or in the part of the tubing that is on the surface.

Other accessories can also be used. Let us mention:

- jet tools with radial and/or end nozzles to get a better cleaning action
- hydraulic motors used with a drilling bit (however, in the case of a small coiled tubing diameter, a large part of the hydraulic power is consumed in the tubing itself before it gets to the motor and little weight can be put on the bit);
- overshots.

These accessories are connected to the tubing either by crimping or by a screwed assembly (Figs. 5.25a and 5.25b).

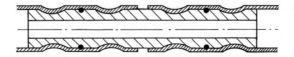

Fig. 5.25a Crimping splice. O ring seal (*Source:* ENSPM Formation Industrie).

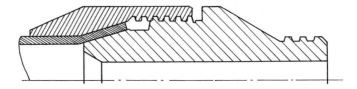

Fig. 5.25b Screwed assembly. Friction cone and metal/metal seal
(*Source:* ENSPM Formation Industrie).

5.3.1.3 Operating considerations

The tubing works in the plastic range at the reel and at the gooseneck on the top of the injector head (to remain in the elastic range with a 1" pipe a curvature radius of approximately 5 m, or 17 ft would be necessary and this is not feasible in practice). As such, pipe fatigue must be taken into account. It is related more to the number of times the pipe is coiled and uncoiled rather than the depth it has worked at. Experience seems to show that the pipe can be coiled and uncoiled at least 20 to 40 times without any excessive risks in routine operations. Acid jobs and other specific conditions would curtail this number. Beyond this point the pipe should be scrapped.

The following operational limits can be mentioned:

- Even if the capacity rating of the unit allows higher speeds, it is advisable to limit speeds to 20 m/min (65 ft/min) when running in, 40 to 50 m/min (130 to 165 ft/min) when pulling out and 0.3 m/min (1 ft/min) penetration rate in sediments (in 5 to 6 m, or 16 to 20 ft stretches at the most).
- As for internal yield pressure, the limit is generally 5000 psi (35 MPa), which allows circulating flow rates of around 80 liters/min (0.5 bpm, barrels per minute), 160 liters/min (1 bpm), and 240 liters/min (1.5 bpm) of fresh water for coiled tubings with a pipe diameter of 1", 1 1/4" and 1 1/2" respectively and a 5000 m (around 16 000 ft) reel. These are relatively small flows that may be too low to raise sediments properly (the required speed is around 30 to 40 m/min, or 100 to 130 ft/min with water). Viscosified water, foam or nitrogen slugs, etc. must then be used.
- As for collapse pressure, the limit is generally 1000 or 2000 psi (approximately 7 or 14 MPa) to take account of the influence of tension on collapse strength.
- As for tension, depending on the diameter and thickness of the pipe the limit corresponds in fact to the weight of about 4000 to 5000 m (13 000 to 16 000 ft) of tubing in air (unless the tubing is composite pipe, i.e. thinner at the bottom and thicker at the top).

To sum up, the coiled tubing can allow circulation in the well, albeit at a moderate flow rate, but it is limited in tensional strength, and, for a small coiled tubing diameter, allows little or no rotation (use of a downhole motor) and does not permit any weight to be set on the downhole tool.

5.3.2 Snubbing

5.3.2.1 Principle and area of application

Snubbing is an older technique (dating back to 1928 in Louisiana) than coiled tubing (around 1960) and yet it developed only moderately for a number of years even in the United States.

Like coiled tubing techniques, snubbing (Fig. 5.26) allows a tubular to be run with a check valve on the end into a live well by means of specialized handling and sealing systems. However, instead of pipe coiled up on a reel, it uses tubing-type pipe lengths run in hole and made up to each other by conventional threaded connections. This means that larger diameter pipe can be used than in the coiled tubing method. Operations naturally remain limited since snubbing pipe has to be run through the tubing set in the well.

The snubbing unit therefore offers better flow capacity, breaking load and rotation capacity and is also able to put weight on the downhole tool. In contrast, tripping takes longer because the lengths of pipe have to be screwed together and due to the procedure for running the connections through the safety stack on the wellhead. Operating this type of unit requires specialized personnel usually consisting of a head of unit and three or four people per shift.

The snubbing unit can perform all of the operations that the coiled tubing unit can, but implementation is longer.

The unit can also perform or make easier some special operations, such as:

- circulate at a higher flow rate (which may offset longer tripping times)
- clean out hard fill and scale that require weight on the tool and rotation
- install a "permanent" concentric tubing to inject an inhibitor, for gas lift, etc.
- spot cement plugs
- do lightweight drilling out (cement plugs, etc.)
- perform some fishing jobs (fish up a wireline or coiled tubing fish, etc.).

5.3.2.2 Snubbing equipment (Fig. 5.27)

A snubbing unit consists basically of the following:

- a pipe handling system
- a wellhead safety system
- a hydraulic power unit

and in addition, the downhole accessories incorporated into the snubbing string.

5. WELL SERVICING AND WORKOVER

Fig. 5.26 Snubbing unit (*Source:* Otis catalog).

5. WELL SERVICING AND WORKOVER

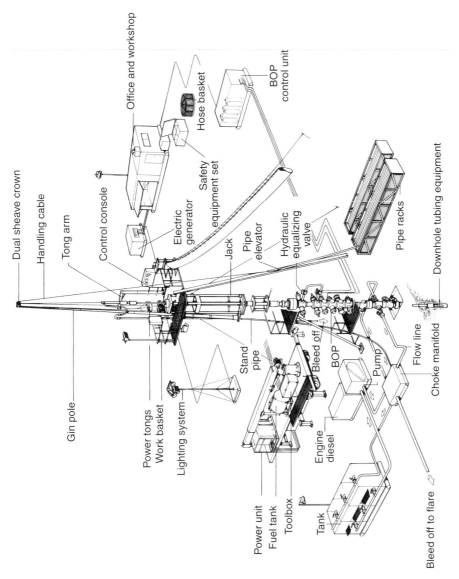

Fig. 5.27 Snubbing unit (*Source*: Flopetrol document).

5. WELL SERVICING AND WORKOVER

A. The pipe handling system

According to the wellhead pressure, the system must be able to push the pipe into the well as long as the pipe weight is not greater than the force exerted by the pressure on the total cross-section of the pipe. Then it must be able to support it.

Hydraulic units are commonly used with double acting jacks equipped with two systems of slips, one stationary and the other mobile (traveling slips). The mobile system, connected to the jack movement, usually only consists of one set of single acting slips (traveling slips). The direction therefore has to be reversed when the balance point (when the weight of the snubbing pipe is equal to the pressure force that is exerted on it) is reached. Note that as long as the pipe has to be pushed into the well (pipe weight lower than the pressure force against it), the operation is said to be in the snub or snubbing phase or the light pipe phase. In contrast when the pipe has to be held up (pipe weight greater than the pressure force against it), the operation is said to be in the strip or stripping phase or the heavy pipe phase.

The stationary system of slips consists of two sets of opposing slips that keep the pipe in place whatever the phase. It is located below the low position of the traveling slips.

With the traveling slips closed and the stationary ones open, the pipe can be tripped over a length corresponding to the stroke of the jacks. Then all that is required to bring the jack back to its original position is to close the stationary slips and open the traveling slips. After the traveling slips have been closed again and the stationary slips have been opened the operation can continue (Fig. 5.28).

An access window can be located below the jack and slip system. It is used to handle any tool that can not be run through the jack and slip system (tubing hanger, drilling bit) with the help of a hanger flange if need be. Getting these tools through the safety system requires a special procedure (see Section 5.3.2.3: Operating considerations).

The hanger flange features dies that prevent any longitudinal or rotational movement of the pipe during special operations (work done on the jack and slip system, etc.). It can have various positions in the stack. Generally it is incorporated into the safety stack, where it must always be located above the safety BOP.

Finally, a rotary table can be added to the unit, usually at the mobile slips. Otherwise a moderate rotation function can be achieved by means of a power swivel-type hydraulic tong.

Besides the cable units (that are found in the United States and need a conventional drilling or workover rig to function), there are two main types of snubbing units:

- long stroke units with a stroke of approximately 11 m (36 ft)
- short stroke units with a stroke of approximately 3 m (10 ft).

Long stroke units (Fig. 5.29) consist of a mast with a travelling block guided by rails and driven by cables actuated by one or two long stroke jacks. They are designed to handle a length of tubing in one go in the strip phase. In contrast, the tubing is not as well guided in the snub phase and this increases buckling problems (see Section 5.3.2.3: Operating considerations).

5. WELL SERVICING AND WORKOVER

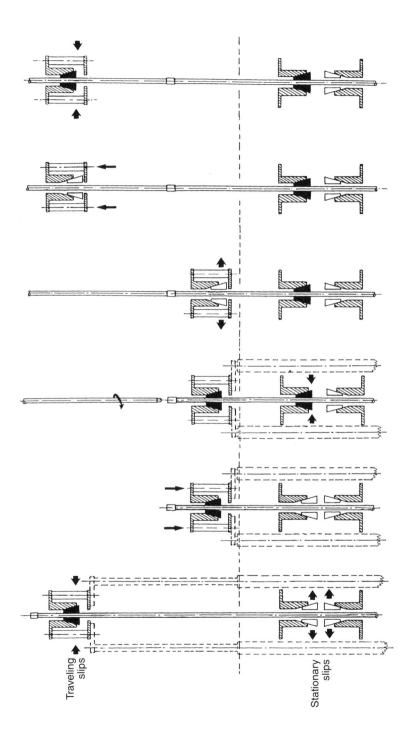

Fig. 5.28 Snub phase running in sequence (*Source*: After a Flopetrol document).

5. WELL SERVICING AND WORKOVER

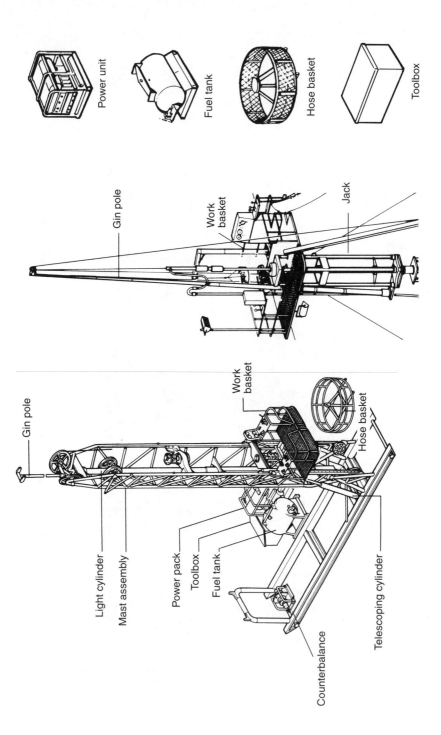

Fig. 5.30 Short stroke unit (*Source:* Flopetrol document).

Fig. 5.29 Long stroke unit (*Source:* Flopetrol document).

Moreover, with long stroke units there is a space problem and offshore there is a problem handling several heavy packages.

Short stroke units (Fig. 5.30) usually have 4 jacks or one concentric jack, and the traveling slips are directly connected to the upper end of the jack(s). Running or pulling a pipe length requires several back and forth movements of the jack(s), but the units are more compact and the tubing is guided better (particularly in concentric jack units where the tubing passes through an open space in the axis of the jack).

There is a very wide capacity range for snubbing units:
- Hoisting capacity in the strip phase goes from 60 000 to 300 000 or even 400 000 pounds (270 to 1330 or even 1780 kN). In the snub phase capacity is usually half that of the strip phase due to jack design.
- As for the diameter of the snubbing pipe, usually at least 3 1/2" and sometimes up to 7 5/8" (respectively 89 and 194 mm) are possible. Actually the snubbing unit can be used not only to work inside the production tubing with a smaller tubular sometimes called a macaroni, but also to pull out the production tubing.

B. The wellhead safety stack

It consists mainly of the following components (from top to bottom):
- a stripper
- a BOP stack with stripping preventers and a safety preventer.

The stripper (Fig. 5.31) provides a seal on the snubbing pipe by means of one, or more commonly two, semi-rigid sealing elements that the pipe slides through. It is used up to wellhead pressures of 1500 to 3000 psi (10.5 to 21 MPa approximately) depending on whether it has one or two sealing elements. It allows tubing connections to pass through while maintaining a seal. Integral joints (integrated in the pipe) are recommended.

When there are two sealing elements, oil is pumped between them at a pressure $P_o = P/2$ (P being the wellhead pressure). As a result:
- each sealing element works at a pressure differential that is half of P
- the pipe is lubricated.

Two ram-type stripping preventers (rams equipped with a wear packing and closing on the snubbing pipe) allow pipe to be run in or pulled out at pressures above and beyond the stripper's working pressure. They are also used for tools that can not pass through the stripper's sealing elements. They are separated by a spacer spool whose height depends on the length of the tools used and on the maximum difference in length between the various pipe lengths to be run in. Figure 5.32 illustrates the BOPs' operating sequence when running a joint into the well. The upper stripping preventer is closed while the pipe is being run in and is opened only to let the joint pass after the lower stripping preventer has been closed. On the contrary, the lower preventer is left open during running in and is closed only for the joint to pass through the other preventer.

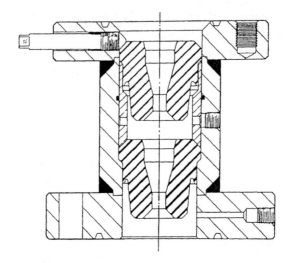

Fig. 5.31 Stripper (*Source: Pétrole et Techniques,* No. 256, October 1978).

The BOPs should be actuated only after equalizing or bleeding off pressure on either side by means of an appropriate mechanism.

A ram-type safety BOP (also closing on the snubbing pipe) completes the wellhead safety equipment. The lowest one in the system, it must remain open during all tripping phases and is used only in static conditions as a safety BOP. It allows the sealing elements of the stripping BOPs and the stripper to be changed in particular, and should never be used to strip or snub.

Other components can be included in the wellhead safety system:

- a hanger flange
- a spherical- or annular-type BOP which is very effective but its packing element can not be changed during an operation
- a shear ram BOP
- a blind ram BOP
- extra stripping or safety BOPs.

C. The hydraulic power unit

A diesel engine drives:

- two main pumps for the jacks
- an auxiliary pump for the BOPs, equalizing and bleed valves, slips, the winch, etc.

D. Downhole accessories

Check valves mounted on the lower end of the pipe are mandatory for snubbing (at least two for safety's sake). If the diameter of the pipe allows for wireline work, instead of using check

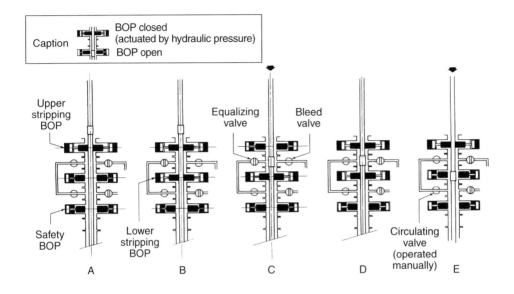

Fig. 5.32 BOP operating sequence when running pipes into the hole (*Source:* After a Flopetrol document).

valves screwed directly onto the tubing, check valves placed in landing nipples screwed onto the pipe can be employed. Once the pipe has been run in and a circulating head and a wireline lubricator have been mounted, the check valves can be retrieved. In this way reverse circulation, etc. can be performed. The check valves will naturally have to be put back into place before the pipe is pulled out if there is any pressure remaining in the well.

One of the check valves is placed far enough from the end of the pipe so that it is above the stationary slips while the pipe shoe is still located under the safety stack. This serves as a warning signal when pulling out, since to get the end of the pipe out of the sealing elements, everything must be closed underneath.

Depending on the operation that is scheduled, different tools can be placed at the end of the pipe: jetting tools, drilling bit with or without a hydraulic motor, fishing tools (overshot, fishing tap, safety joint, jar, etc.).

5.3.2.3 Operating considerations

When tripping begins (during the snub phase) the pipehead is compressed due to the well pressure, with the resulting risk of buckling. Consequently, the length of unguided pipe (i.e. generally from the stripper to the traveling slips) must be limited to less than the critical buckling length. Otherwise the pipe will buckle and may be ejected from the well. This situation, which depends on the wellhead pressure and on the pipe diameter, is at a maximum at the beginning of the running in phase or at the end of the pulling-out phase (no pipe weight to offset the pressure). For example at the beginning of the running in phase with a wellhead pressure of 10 MPa (1450 psi) the critical length is:

- about 4 m (13 ft) for a 1.9" (48 mm) tubing
- about 7 m (23 ft) for a 3 1/2" (89 mm) tubing.

When the traveling system has only one set of single acting slips, it will have to be reversed at the balance point. In order to go straight from a status where slips are needed in one direction to one where they are needed in the opposite direction, it is possible to:

- start with empty pipe and fill it to jump over the balance point when running in
- use the wellhead pressure both when running in and pulling out, i.e. get the well to flow or close it depending whether the rest of the trip is made with the well closed or flowing.

The lifetime of sealing elements is highly variable. Stripper sealing elements can just as easily allow a round trip to 3000 m (10 000 ft) without any problem as need to be changed every 80 meters (260 ft) when pulling out a rough pipe, covered with asphalt deposits, etc. When the trip is made using a ram-type stripping BOP, the packing element is changed every 20 to 30 pipe lengths. Meanwhile a spherical BOP can sometimes hold up for the whole operation with no trouble.

Tripping speeds depend on whether the pipe is going through the stripper alone or through the ram-type stripping BOPs. In the first case speeds of approximately 600 m (2000 ft) per hour can be reached (i.e. 10 m/min, or 30 ft/min in comparison with the 20 to 40 m/min, or 60 to 120 ft/min in coiled tubing operations depending on whether the tubing is being run in or pulled out). In the second case the speed only reaches one-quarter of this figure, i.e. 150 meters (500 ft) per hour.

In principle, no snubbing operations are done during the night, the working day starts at dawn and ends one hour before nightfall. The BOPs are then closed, bled off and locked along with the slips. A closed valve is screwed onto the tubing. If the operation does have to continue after nightfall it should last the shortest time possible and the site should be well lit. However, the present trend seems to be toward night work and this means adequate lighting and some restrictions on the operations that are authorized. In particular, no tubing tripping is done in the snub phase during the night.

5.4 OPERATIONS ON KILLED WELLS

For some types of servicing jobs, especially when the tubing and its equipment have to be pulled out, it may be preferable or necessary to kill the well beforehand, i.e. place a control fluid in the well that exerts a hydrostatic pressure greater than the reservoir pressure. Work can then be done with the well open without any wellhead pressure.

Generally speaking, the purpose is to modify the completion configuration and the techniques used are exactly the same as during initial completion. However, particular care should be taken to control the well properly and to redefine the new configuration to suit present conditions (which can be very different from initial conditions) and future conditions (which have become easier to anticipate) as well as possible. This comment does not just apply to completion that is rehauled to adapt to a different well purpose, but to all workover jobs.

5.4.1 Means of acting on killed wells

The required means depend mainly on:
- the depth of the well
- the equipment already installed in the well
- the job that needs to be done.

Lightweight units, called servicing or pulling units, can be used. These rigs are mobile, lightweight, readily installed over the wellhead and principally designed to pull out or run in pumping rods or tubings, usually at depths of less than 2000 to 2500 m (6500 to 8000 ft). In their simplest form they can be no more than cranes.

Larger units, comparable to a certain extent to drilling rigs, are also used and are called workover rigs. They can be lightweight, medium or heavyweight.

The unit is chosen in relation to the planned operation depending on its technical specifications (hoisting capacity, possibility of rotation, pumping capacity, safety and ancillary equipment, etc.), on its daily cost and geographical availability. In practice priority often unfortunately goes first to geographical availability and then to daily cost. This does not necessarily prove to be the most economical in comparison with the overall cost of the operation (duration, results, etc.).

Whatever the type of unit, it must have the appropriate specialized equipment to be able to complete the job in the safest and most efficient way. The required equipment includes the following in particular:

- safety equipment: BOP, BPV, gray valve, etc.
- high pressure pumps, storage tanks, etc.
- hoisting, pipe makeup and fishing equipment suitable for the small diameter drill pipe and tubing that are used in workovers
- wireline equipment (including the corresponding fishing equipment), and even well logging equipment.

5.4.2 General procedure of an operation

The phases and sequencing of an operation vary of course from one job to another. They mainly depend on the equipment already installed in the well and the condition it is in, on what needs to be done and how the operation is going to proceed in practical terms. However, the detailed steps discussed below are generally involved.

5.4.2.1 Preparing the well (before the servicing or workover unit arrives)

This mainly means:

- checking the status of the well by wireline techniques (checking the tubing, tag sediment), and sometimes
- checking well integrity (pressure testing, etc.)
- opening a circulating device downhole.

5.4.2.2 Putting the well under provisional safe conditions (before rigging up the servicing or workover unit)

This safety operation in fact also involves all the nearby wells (particularly if the well is in a cluster) that might be hit when the servicing unit is getting set up. It consists in setting plugs in the tubing in order to install the servicing unit on the wellhead under optimum safe conditions.

There are three basic ways of doing this:

- using plugs run by wireline and locked in landing nipples in the tubing (generally at the bottom of the well, near the packer)
- closing the subsurface safety valve if there is one
- setting a back pressure valve in the tubing hanger.

At least two of these safety barriers are normally used.

The various lines connected to the wellhead on the surface (flowlines, etc.) also have to be isolated and dismantled and nearby equipment that might be damaged has to be bled off.

5.4.2.3 Installing the servicing or workover unit

Once the well has been placed under provisional safe conditions, the rig and all its equipment (tank, pumps, workshop, etc.) can be set up in accordance with safety distances, rules and regulations. However, the Christmas tree is not yet removed to be replaced by the BOPs.

5.4.2.4 Killing the well

A well is considered to be perfectly killed when the workover fluid, whose specific gravity is appropriate for the reservoir pressure, totally fills up the well (i.e. the inside of the tubing, the tubing-casing annulus and the space under the packer).

The workover fluid is prepared in sufficient amounts (three times the total volume of the well). In actual fact the workover fluid is just a completion fluid, since the same properties are required of it, mainly:

- to keep the well under control by its hydrostatic pressure
- to carry up cuttings if drilling out or milling is planned
- not to damage the formation
- no fluid loss in the formation.

After the plugs that were set in the tubing to allow the rig to be installed on the wellhead have been retrieved, the workover fluid is displaced into the well either by circulating or by squeezing. Then the well's stability is observed. In some cases killing is carried out before the rig is installed. Killing techniques as such are dealt with in section 5.4.3.

5.4.2.5 Replacing the Christmas tree with the BOPs

Since the workover fluid is keeping the well stable, only one mechanical safety barrier is deemed necessary (preferably the downhole plug and/or the SCSSV and/or the BPV).

The Christmas tree can then be dismantled at the tubing-head spool and replaced by BOPs which will be tested. This operation should be completed as quickly as possible. As a result, the personnel must be mobilized, all the equipment ready, appropriate handling and hoisting equipment available, the wellhead bolts checked, etc.

5.4.2.6 Removing completion equipment

Pulling out the downhole equipment can then start, after the BPV (or other plugs that may have been set in the well) has been retrieved. If there is a kick while tripping, it is necessary to be able to close quickly not only the annulus (with the BOP tubing rams) but also the tubing itself. The corresponding safety device (gray valve, etc.) must be on the workover rig floor ready for use (thread compatible with tubing thread, manoeuvring wrench, etc.).

The procedure as such for removing the downhole equipment depends on the type of equipment and its condition. Particularly important is the type of packer, retrievable or permanent, and if permanent, the type of connection between the tubing and the packer (seal alone or seal plus anchor). With a retrievable packer, and especially if there are any doubts as to the condition of the tubing, it is better not to attempt to unseat the packer by pulling directly on the tubing. Instead, it is often wise to cut off the tubing a few meters above the packer (by means of an explosive charge run on an electric cable), then to run in drill pipe equipped with an overshot (see Section 5.4.5) to unseat the packer.

Furthermore, whatever the killing method, there is always a volume of oil and/or gas trapped under the packer. It is important to circulate it out as soon as possible (for example after unsetting the retrievable packer or after disconnecting the tubing-packer seal if the packer is permanent).

Whenever tripping out, care must be taken to avoid swabbing (particularly when the packer is pulled out) and to keep the well full (offset the volume of tubing steel by an equal volume of workover fluid). Likewise, the well's stability must frequently be checked.

5.4.2.7 Downhole operations, recompletion, replacing BOPs by the Christmas tree and start up

The techniques are the same as those used when the well was originally completed and so they will not be developed here. Note, however, that the bottomhole is usually checked beforehand by running in a drill string equipped with a drill bit and a scraper.

5.4.2.8 Moving out the servicing or workover unit

In the same way as for initial completion, the rig can be moved out before or after the well is brought back on stream. The same safety rules and regulations are complied with (especially setting plugs).

5.4.3 Considerations on killing the well

The workover fluid used to kill the well is usually displaced either by circulating or by squeezing.

5.4.3.1 Killing by circulating

Except for special cases, circulating is normally the preferred way to kill the well. The fluid is circulated as deep as possible in the tubing (above the packer, however):
- either through a circulating device operated by wireline
- or through a perforation usually achieved with the electric cable (shaped charges).

A reverse circulation is usually the option chosen for gas wells or high GOR oil wells, in other words the fluid is pumped into the annulus with returns through the tubing. In this way the oil and gas are displaced more thoroughly. With direct circulation the lighter oil and gas tend to migrate in the heavier workover fluid which is located above them (with reverse circulation the workover fluid comes up under the oil and gas). Accordingly, there is no certainty as to the specific gravity of the workover fluid that is left in place in the well, hence as to the actual killing of the well. However, after a volume corresponding to the tubing volume has been pumped in through the annulus, and therefore after the oil and gas have been pumped out of the well, direct circulation is generally used to homogenize the fluids present in the tubing and annulus.

In contrast, since annular pressure losses are generally lower than tubing pressure losses (for oil wells), with direct circulation the overpressure exerted on the pay zone by pressure losses is minimized (hence problems of lost circulation and plugging are also minimized).

Whether circulation is reverse or totally direct, the choke is applied on the returns to keep the bottomhole pressure higher than the reservoir pressure all throughout the operation, thereby preventing any kicks.

When circulation is finished, not all of the fluids located below the circulating device have been replaced by killing fluid, so killing is only partial. If the injectivity is not too bad killing can be improved after circulation by squeezing the oil and gas still remaining from the circulating device to the perforations. However, some oil and gas will remain trapped in the annulus under the packer opposite the tubing tail pipe.

Sometimes the workover fluid is circulated after setting a plug in the lower end of the tubing. As a result there is no risk of kicks or lost circulation into the formation during circulation, but there is no certainty that the plug can be retrieved, especially if it gets covered with deposits.

5.4.3.2 Killing by squeezing

The squeeze technique is used:
- when the circulating method can not be used (tubing has got a hole in it "near" the surface, wireline is impossible because of a collapsed tubing or a fish, volume under the tubing shoe is too large, etc.); or
- when injectivity is very good (true especially for gas wells).

Squeezing is done only after an injectivity test to be sure that injectivity is sufficient. This is because of the same drawback as with direct circulation (migration of oil and gas in the killing fluid), greatly compounded by the fact that the flow rate possible with a squeeze job is often much lower than with circulation.

If the injectivity test is unsatisfactory it is necessary to:
- modify the squeeze conditions, for example:
  - squeeze at a lower flow rate, provided the total injected volume is increased (be careful to maintain a sufficient flow rate however, because of migration)
  - for a gas well, perform a series of squeezes and bleed offs at constant bottomhole pressure to allow the gas to migrate up to the surface without fracturing the formation

- accept to squeeze at a fracturing rate
- before squeezing, circulate at the lowest depth possible (that can be reached for perforating, etc.)
- or revise the program thoroughly and use a technique such as coiled tubing or snubbing just to kill the well or to carry out the complete operation.

This means that the injectivity test has to be performed long before the workover is done. The choice of means to be utilized and the workover program design both depend on the injectivity test.

In the same way as for circulation, here too killing is not complete when the squeeze is over. Oil and gas trapped in the annulus under the packer as well as all of the annular fluid still remain in the well.

5.4.3.3 Observing the well

After killing by circulation or squeeze the well must be observed to check that:
- there is no wellhead pressure
- the levels are stable
- there is no gas cutting on the surface.

Whatever the observation time, there is no absolute guarantee that the well is stable. The time depends on the method used to kill the well, on the way the operation actually proceeded and on outside signs of phenomena such as thermal expansion. It can vary from one to several hours.

If the well should prove unstable, killing must be started all over again either with the same method (possibly modifying the characteristics of the control fluid or using spacer fluids, etc.) or with another method. Once again the well will be observed and so on and so forth until it remains perfectly stable.

5.4.3.4 Final killing phase

We have seen that, whatever the method used, there is always a certain volume left in the well where it is impossible to place the workover fluid. This volume must be circulated as soon as the completion equipment removal operations allow.

5.4.4 Depleted reservoirs

Depleted reservoirs pose particular problems as soon as their reservoir pressure is such that conventional workover fluids (brines, oil base fluids) exert a considerable overpressure.

Then the following problems are posed:
- losses and/or formation damage; and
- well kick off after operations.

5. WELL SERVICING AND WORKOVER

5.4.4.1 Losses and formation damage

The first solution is to use special light fluids, but besides foam (specific gravity from 0.1 to 0.2) there is as yet no fluid readily available on the market in the neighborhood of the 0.2 to 0.8 range (between foam and oil base fluids).

The second solution is to protect the formation with "temporary" blocking agents. These are usually products that are unstable with temperature or highly soluble in acid. The problem with them is that they are never 100% destructible.

The third solution consists in accepting to work with a "lost" level, i.e. with the level somewhere in the well and not any more on the surface. However this causes safety problems (it is hard to check well stability constantly) and is incompatible with operations that require circulation (cleaning, etc.).

The fourth solution, when only the production string above the packer is involved, consists in using equipment installed for this purpose during the original completion operation. It allows the tubing alone to be pulled out leaving the packer and a downhole plug in an extension in place in the well. Figure 5.33 shows this type of equipment and illustrates a procedure that can be used in this case.

The fifth solution might be to use a unit allowing work on a pressurised well, such as coiled tubing or snubbing if they are appropriate for the job (cleaning the bottom of the well, etc.).

5.4.4.2 Start up problems

The techniques described in initial completion are applicable at this stage, except that there is usually no available kick off fluid that is light enough. In some cases the problem does not crop up because the purpose of the operation was to install an artificial lift system or readapt the old one. In the other cases it is very common to use a coiled tubing unit for nitrogen start up or temporary gas lift or, failing that, to use a wireline unit for swabbing.

5.4.5 Fishing tools

Special tools, called fishing tools, are necessary to carry out certain phases in retrieving the completion equipment or for a fishing operation (due to a "bungled" job that left a fish in the well). There are a large number of varied tools so we will restrict our discussion to mentioning the main types (for more details see documentation on drilling or on fishing during drilling).

Depending on the problem that has arisen and the information available, and possibly after having run in an impression block, the following can be used, among others (Fig. 5.34 first or second part):

- external catch tools such as:
  - die collars (Fig. 5.34a), overshots (Fig. 5.34b) to retrieve a tubular

5. WELL SERVICING AND WORKOVER

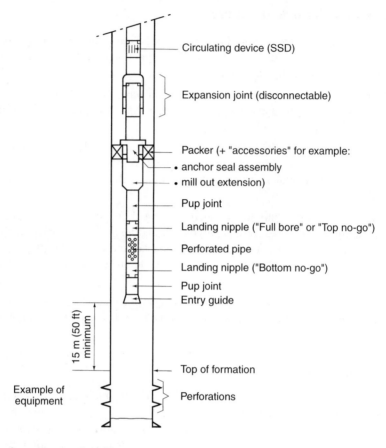

Procedure for pulling out the tubing:
1. A plug is set in the landing nipple of the fixed part of the disconnectable joint (beforehand a plug can also be set in the top no-go landing nipple).
2. The circulating device is opened or the production string is perforated just above the disconnectable joint.
3. The workover fluid is circulated into the well.
4. A BPV is set in the tubing hanger.
5. The Christmas tree is removed.
6. The BOPs are installed.
7. The BPV is replaced by a TWCV.
8. The BOPs are tested (between the TWCV and the blind rams); a tubing is screwed into the tubing hanger and the pipe-ram BOPs are tested against the tubing.
9. The tubing hanger is unlocked (tie-down screws).
10. The upper part of the production string is pulled out.

Fig. 5.33 Example of downhole equipment designed for partial retrieval of the production string (*Source:* ENSPM Formation Industrie).

5. WELL SERVICING AND WORKOVER

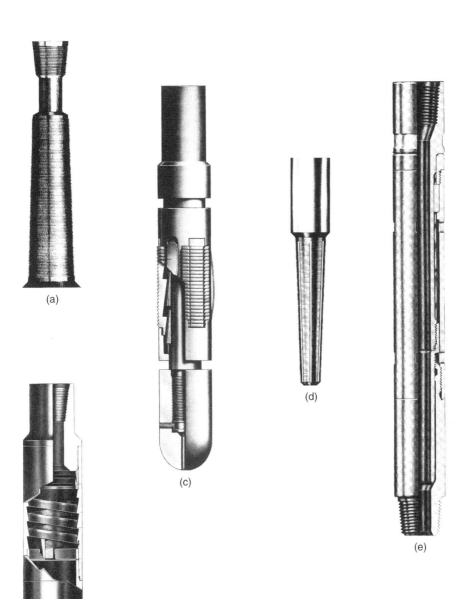

Fig. 5.34 (Part 1) Fishing tools (*Source:* Servco catalog).
a. Die collar. b. Overshot. c. Releasing spear. d. Taper tap.
e. Bumper sub.

5. WELL SERVICING AND WORKOVER

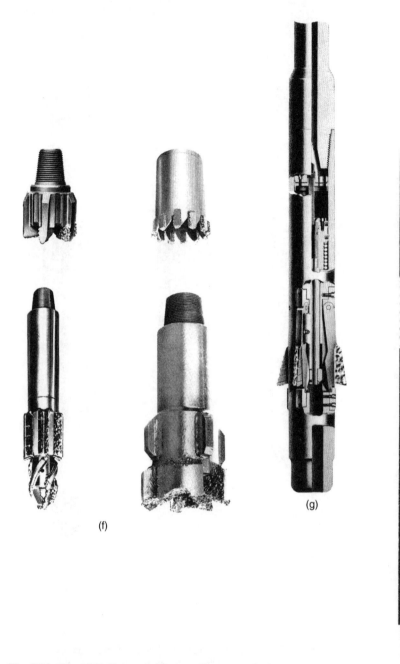

Fig. 5.34 (Part 2) Fishing tools (*Source:* Servco catalog).
f. Milling tool. g. Section mill. h. Safety joint.

- junk catchers to collect small pieces of metal
- internal catch tools such as releasing spears (Fig. 5.34c) or taper taps (Fig. 5.34d)
- jarring tools such as mechanical bumper subs (Fig. 5.34e) or hydraulic jars that can have added jar accelerators or intensifiers
- milling or washover tools (Fig. 5.34f)
- tools to unscrew and cut tubing or casing such as, on the one hand, reversing tools, left-hand thread tap or the back off shooting technique, on the other hand, section mills for mechanical cutting (Fig. 5.34g), jetting tools for hydraulic cutting or explosive cutters
- safety joints (Fig. 5.34h) allowing the workover string to be disconnected at the joint depth if the fish will not come out.

5.5 SPECIAL CASES

5.5.1 Operations on horizontal wells

Except for wireline, all the conventional servicing and workover methods described previously can be used on horizontal wells. There are sometimes parameters that complicate the job such as the effect of gravity or the length of the horizontal drain hole. They may require adaptations (better centering, etc.) or may restrict the type of operations that can be contemplated.

Techniques other than wireline must be used to take measurements in the horizontal drain hole. Wireline is not suited to horizontal wells, even though it can be used in deviations up to approximately 65 to 70°, or even 80° with roller mounted tools.

First of all, the "Measurements While Drilling" (MWD) technique. The logging tool is incorporated in the drill string and the measurement is transmitted to the surface as drilling progresses. Information is in the form of pressure pulses transiting through the drilling fluid (or via an electromagnetic system). Conventional tools are now available (gamma ray, resistivity, neutron, etc. logs) and research and development efforts are under way in this area.

Another technique associates the drill pipe and a conventional electric cable. This is the Simphor (Système d'Information et de Mesures en Puits HORizontaux) developed by Elf and the Institut Français du Pétrole, and the same process has also been developed under other names by service companies (Fig. 5.35):

- conventional electric tools are fixed to the end of the drill string by a connecting sub that allows the electric cable to be connected later on
- the drill pipe is run in to the beginning of the zone where the logging is going to be done
- a side port (with a sealing system) for the electric cable to pass through is then screwed onto the top of the drill string

- a "locomotive" equipped with cups and fixed to the electric cable is pumped through the pipe pulling the cable down to the downhole connector
- logging can be done running in and then pulling out by adding or removing extra lengths of pipe.

Note that both MWD and Simphor can only perform measurements before the well is completed.

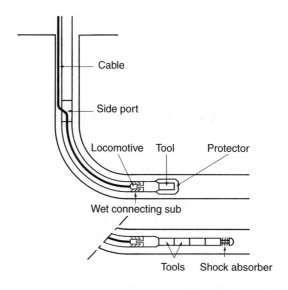

Fig. 5.35 Simphor process (*Source: J. of Petroleum Technology,* October 1988).

The pump down stinger technique can be used either through the drill pipe or through the tubing. The drill pipe or tubing is in place and stops before the depth where measurements are going to be made (Fig. 5.36):

- The logging tools are screwed onto the lower end of an extension consisting of precabled pipe giving mechanical and electric continuity. The extension is long enough to extend the drill pipe or tubing beyond the zone that is going to be logged.
- The upper end of the extension pipe is connected to the electric cable by a locomotive.
- Pumped through the pipe or tubing, the locomotive pushes the tools out of the tubular (depending on the length of the extension) carrying the electric cable along with it.
- The tools are retrieved by pulling on the cable.

The pump down stinger technique is currently limited to small diameter tools. Moreover, unless the injectivity is sufficient, completion must be without a packer or with two tubings to allow the return flow of the pumped fluid. It can be applied to tools other than measurement tools in a horizontal drain hole and in this case the cable can be eliminated. However, reverse circulation is then mandatory to retrieve the locomotive (completion without a packer or with two tubings).

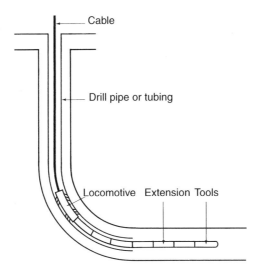

Fig. 5.36 Pump down stinger process (*Source: J. of Petroleum Technology,* October 1988).

Another technique is to use a coiled tubing unit to push the tools down into the well. Here the electric cable is run into the coiled tubing before it is rolled up on the reel and the tool is connected directly to the end of the coiled tubing. The electric signal is picked up on the surface via a rotary electric connector on the reel axis. The method is fast and easy to implement, but limited as to the capacity to push tools. It is therefore not recommended in an open hole. In addition, it allows only limited horizontal movements (approximately 200 m, or 650 ft) when standard diameter rather than small diameter tools are used (such as production logging tools).

The various techniques presented above are not restricted in application to horizontal wells. They can also be used in highly deviated or difficult wells (very viscous fluid, severe dog leg, i.e. sudden variations in deviation in one direction and then the other, etc.) where wireline techniques can not.

5.5.2 Operations on subsea wells

Operations on subsea wells must be exceptions as the cost of the job is disproportionately higher than for a wellhead above water. This is why subsea wellheads were long reserved for extensions of satellite fields where fixed structures were not warranted and production could be stopped if there was any trouble (to wait for a favorable weather window or for a servicing rig to be available, etc.).

In order to cut down on operations and make them easier, the present trend is:
- For all equipment, from downhole equipment to wellhead control systems, to evolve toward systems that are as simple as possible instead of complex.
- To anticipate how well conditions will change and what future needs will be in order to extend the time between completion and workover, or workover and workover.
- To give priority to installing wet wellheads (in contact with the marine environment) rather than dry wellheads (placed in a watertight housing at atmospheric pressure). Dry wellheads were installed mainly in the beginnings of subsea production and today account for less than 2% of the subsea wellheads now in place in the world.
- To move toward systems that do not require divers for installation and maintenance. However, installations located in water depths that are accessible to divers must be designed so that divers can work on them without running any risks (Among the 1117 subsea trees in the world in 1996, 867, i.e. 78%, are located in the 0 to 200 m, or 0 to 650 ft range; 127, i.e. 11%, in the 200 to 400 m, or 650 to 1300 ft range; and 123, i.e. 11%, in water depths over 400 m or 1300 ft. In 1996, 1041 subsea trees were scheduled for installation between 1997 and 2001, among which 339 in water depths between 200 and 400 m, or 650 and 1300 ft, and 243 in water depths greater than 400 m or 1300 ft).
- To equip wellheads with visual displays and manual emergency controls for all connectors and valves so that observation and action are easier for divers or Remotely Operated Vehicles, ROV.

Servicing on the wellhead is carried out:
- either under water by ROV or divers, or
- after bringing all or part of the Christmas tree up to the surface.

Depending on the operation and on local conditions, this type of servicing requires a floating support vessel that may vary in size.

For measurement and maintenance operations, the basic technique is wireline. Various methods are used:
- A drill ship, a semi-submersible platform or a jack up platform for shallow or medium water depths, and wireline operation from the support unit after installing a servicing riser (with a lubricator), said riser being supported by the unit.
- A permanent rigid riser tensioned by a floating buoy, the wireline winch installed on the buoy with a service boat supplying the hydraulic power.
- A permanent submerged riser tensioned by a submerged buoy (less than 15 m, or 50 ft approximately). When servicing is required a lightweight extension is installed (end of the riser and lubricator) and work is done from the service boat.
- A submerged lubricator and automatic winch directly connected to the subsea wellhead and hydraulically controlled from the surface from a service boat.

Some subsea wellheads have been equipped for the Through Flow Line technique, TFL. Here the tools are run in and pulled out of the well by pumping them through the flowline from a "nearby" production platform by means of a locomotive equipped with sealing elements rather than by wireline. It requires specific completion equipment, particularly in order to pump the tools down and back up again (Fig. 5.37):

- a double flowline
- a TFL wellhead, a double wellhead with loops to connect up with the flowlines
- a double tubing connected in the lower part by a circulating sub.

Workover requires remobilizing a heavyweight support unit (drill ship, semi-submersible or possibly jack up).

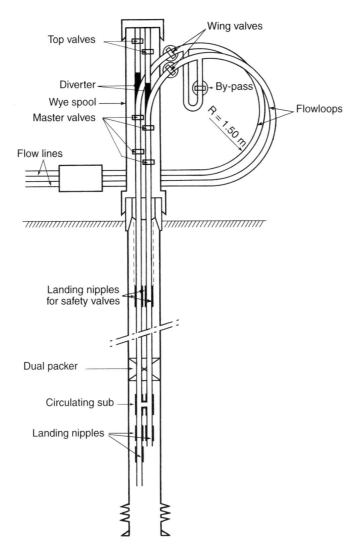

Fig. 5.37 TFL equipment (one producing zone but two tubings) (*Source:* ENSPM Formation Industrie).

BIBLIOGRAPHY

Chapter 1
INTRODUCTION TO COMPLETION

KRUEGER R.F., "An overview of formation damage and well productivity in oilfield operations", *J.P.T.,* vol. 38, n° 2, 1986.

Le matériel de complétion, Total CFP, Cours à l'ENSPM, 1987.

PATTON L.D., ABBOT W.A., "Well completions and workovers: the system approach", *Energy publications,* 1985.

PERRIN D., *Information "completion",* Cours à l'ENSPM, réf. 29 255, 1981.

VEILLON D., *Bases techniques de la complétion. Tome I : Notions de base et liaison couche-trou,* Cours à l'ENSPM, réf. 35 654, 1987.

VEILLON D., *Bases techniques de la complétion. Tome II : Traitement des réservoirs,* Cours à l'ENSPM, réf. 35 865, 1988.

Divers documents de cours ENSPM Formation Industrie.

Divers documents de fabricants ou sociétés de service : Baker, Servco.

Chapter 2
CONNECTING THE PAY ZONE
AND THE BOREHOLE

BOUTECA M., SARDA J.P., "État de l'art en fracturation hydraulique", *Revue Inst. Franç. du Pétrole,* vol. 42, n° 1, 1987.

The essential of perforating, Schlumberger.

Fluid selection guide for matrix treatments, Prepared by Bertaux J., Edited by Piot B. and Thomas R., Dowell Schlumberger, 1986.

PATTON L.D., ABBOT W.A., "Well completions and workovers: the systems approach", *Energy publications,* 1985.

MONTIGNY O., de COMBE J., "Horizontal well operation – 3. Hole benefits, reservoir types key to profit", *Oil and Gas,* vol. 86, n° 15, 1988.

PERRIN D., *Information "completion",* Cours à l'ENSPM, réf. 29 255, 1981.

SPREUX A., GEORGES C., LESSI J., "Horizontal well operation – 6. Most problems in horizontal completions are resolved", *Oil and Gas,* vol. 86, n° 24, 1988.

VEILLON D., *Bases techniques de la complétion. Tome I : Notions de base et liaison couche-trou,* Cours à l'ENSPM, réf. 35 654, 1987.

VEILLON D., *Bases techniques de la complétion. Tome II : Traitements des réservoirs,* Cours à l'ENSPM, réf. 35 865, 1988.

Divers documents de cours ENSPM Formation Industrie.

Divers documents de fabricants ou sociétés de service : Baker, Dowell, Flopetrol, Schlumberger, Howard Smith.

Chapter 3
THE EQUIPMENT OF NATURALLY FLOWING WELLS

AVIGNON B., *Procédures de complétion,* Support de cours à l'ENSPM Formation Industrie, 1986.

GABOLDE G., NGUYEN J.P., *Formulaire du foreur,* Éditions Technip, Paris, 1989.

MARION G., *Le travail au câble dans les puits en production (tome 2),* Cours ENSPM, réf. 10 404 (texte) et 10 404 A (planche), 1964.

ORIEUX P., *Équipement des puits éruptifs,* Éditions Technip, Paris, 1966.

Complétion et reconditionnement des puits, Programmes et modes opératoires. CSRPPGN, Comité des Techniciens, Éditions Technip, Paris, 1986.

Formulaire du producteur, Éditions Technip, Paris, 1970.

Le matériel de complétion, Total CFP, Cours à l'ENSPM Formation Industrie, 1988.

"World oil's 1989 tubing tables", *World oil,* January 1989.

Divers documents de cours ENSPM Formation Industrie.

Divers documents de fabricants et de sociétés de service : AVA, Baker, Camco, Cameron, Flopetrol, FMC, Otis, Schlumberger, Vallourec.

Chapter 4
ARTIFICIAL LIFT

Activation des puits, Séminaire ENSPM Formation Industrie, "Gas lift", Conférence présentée par Bertain A., 1987.

Activation des puits, Séminaire ENSPM Formation Industrie, "Comparaison des moyens d'activation" Conférence présentée par Jacquart J., 1988.

Activation des puits, Séminaire ENSPM Formation Industrie, "Le pompage par tiges", Documents de Laulanie M. H. et Parpant J. présentés par Miquel M., 1988.

Activation par pompage, Séminaire ENSPM Formation Industrie, "Puits en pompage : méthode et exemples de calculs", Conférence présentée par Miquel M., 1990.

CORTEVILLE J., HOFFMANN F., VALENTIN E., "Activation des puits : critères de sélection des procédés", *Revue Inst. Franç. du Pétrole,* volume 41, n° 6, Éditions Technip, Paris, 1986.

COUPIN A., "Le pompage centrifuge", *Activation des puits,* Séminaire ENSPM Formation Industrie, réf. 25 864, février 1978.

DEMOULIN P., "Le pompage hydraulique", *Revue "Forage",* n° 100, 1983.

Formulaire du producteur, Éditions Technip, Paris, 1970.

de SAINT-PALAIS J., FRANC J., *Cours de gas lift,* Éditions Technip, Paris, 1979.

Documents API :

- SPEC. API RP 11 L, February 1977, 3rd edition, API Recommended Practice for Design Calculations for Sucker Rod Pumping Systems (Conventional Units).
- Vocational Training Series : Book 6, 1984, Gas lift.

Divers documents de fabricants et sociétés de service :

- Gas lift : Camco, Flopetrol, Mas Murry, Macco Schlumberger, Otis, Teledyne Merla.
- Pompage : Axelson, Centrilift, Deltax, Harbison Fisher, Guiberson, Keystone, Kobe, Leutert, Lufkin, Mape, PCM, Reda, Trico Industries, Weir Pump Ltd.

Chapter 5
WELL SERVICING AND WORKOVER

"Le coiled tubing", *Intervention par tubing concentrique,* Conférence présentée au Séminaire ENSPM Formation Industrie par Dowell, 1979.

FRATACCI P., *Les reprises de puits,* Support de cours à l'ENSPM Formation Industrie, 1986.

GAILLOT G., *L'équipement de surface en travail au câble,* Éditions Technip, Paris, 1984.

MARION G., *Le travail au câble dans les puits en production* (Tome II), Cours ENSPM, réf. 10404 (texte) et 10404 A (planches), 1964.

"Le snubbing", *Intervention par tubing concentrique,* Conférence présentée au séminaire ENSPM Formation Industrie par Flopetrol, 1979.

SOREL R., *Cours de production. Tome XII : L'entretien des puits en cours d'exploitation,* Éditions Technip, Paris, 1965.

SPREUX A.M., LOUIS A., BROCCA M., "Logging horizontal wells: field pratice for various techniques", *J.P.T.,* octobre 1988.

TINCHON J., "Le coiled tubing", *Bulletin "Forage",* n° 120, 1988.

TUECH M., CANO F., "Le reconditionnement des puits par unité de workover hydraulique (snubbing)", *Pétrole et techniques,* n° 256, 1978.

Divers documents de cours ENSPM Formation Industrie.

Divers documents de fabricants ou sociétés de service : Flopetrol, Otis, Servco.

INDEX

Numbers **in bold type** refer to pages where the word appears in a paragraph title

A

absolute permeability
  (*refer also to* permeability), **78**
accelerator
  jar accelerator, 303
accessorie
  downhole accessories (*refer also to* downhole equipment), **281, 290**
acid, 91
  acid frac, 97
  acid washing, **75**
  ARC(Acid Response Curve), 92
acidizing, 91, 113
additive, 43, 93, 99
afterflush (*refer to* overflush)
agent
  antislude agent, 93
  complexing (*or* sequestring) agent, 93
  diverting agent, 94, 99
  gelling agent, 99
  propping agent, 97, **99**
  sequestring (*or* complexing) agent, 93
  temporary plugging agent, 94, 299
  wetting agent, 93
alternate selective completion
  (*refer also to* completion), **30**
ambient safety valve
  (*or* pressure operated safety valve), 155
American Petroleum Institute
  (*refer to* API)
anchor,
  gas anchor, **178**
  seal assembly, **135**
  tubing anchor, 159, **178**
annular fluid (*or* packer fluid), **131, 172**
antisludge agent, 93
API (American Petroleum Institute), **60**, 120, 122, **124, 126**, 180, 190, 262

appraisal well, 3, 12
ARC (Acid Response Curve), 92
artificial lift, 7, 10, 19, **175**
  choosing an artificial lift process, **247**
assisted drive, 7

B

B unit, 190
back surging, 74
back-off shooting, 303
Back-Pressure Valve (*refer to* BPV)
ball sealer, 94
Basic Sediment and Water (*refer to* BSW)
beam
  walking beam unit, **188**
bean
  choke bean, 118
bellows valve, 233, 234
big hole charge, 65
blast joint, 117, 153
blind ram (*refer also to* ram), 281, 290
BlowOut Preventer (*refer to* BOP)
body
  body yield strength, 126
  body yield stress, 124
BOP (BlowOut Preventer), **37**, 161, **164** 170, 259, **264**, 280, 289, 294, **295, 296**
  safety BOP, 290
  stripping BOP, 289, 292
borehole
  checking the borehole, **33,** 163
  pay zone-borehole connection, **25, 33, 37,** 110, **164, 256,** 257
bottom no-go or **no-go** (*refer to* landing nipple)
bottomhole, 255, 256
  bottomhole pressure P_{BH}, 13, **17**
  bottomhole pump, **180**

INDEX

box
    stuffing box, **263**
BPV (Back-Pressure Valve), 118, 171, 294
breakdown pressure (*or* fracturing pressure), 97
breaker, 99, 104
breathing, 130, 178
bridge plug, 51, **56**
BSW (Basic Sediment and Water), 105
buckling, 178
bumper sub, 303
burst pressure (*or* internal yield pressure), 126

C

C (empirical equation Coefficient), 15, **20**
C unit, 190
cable
    wireline cable, 262
cake
    filter cake, 38
caliper, 272
cap
    gas cap, 7, 20, 107
    tree cap, 116
cased hole,
    cased hole completion (*refer also to* completion), 25, **26**, **84**, **87**
    cased hole logging, **163**
casing, 26, 27, **43**, 117, **163**, 254
    casing diameter, **43**
    casing gun, 66, 164
    casing head spool, 118, 161
    casing operated valve (*refer also to* gas lift), 233
    drilling and casing program, **5**
    fixed casing pump, 201
    free casing pump, 201
    production casing, **6**
catch tool, 299, 303
cave, 84
CBL-VDL (Cement Bond Log - Variable Density Log), **46**
CCL (Casing Collar Locator), 33, 73, 163, 165
cement
    cement job (*see below*)
    cement retainer, 51, 53, 55

cement job (*refer also to* cementing)
    evaluating the cement job, **44**
    restoring the cement job (*refer also to* remedial cementing), **44**
cementing (*refer also to* cement), **6**, **43**, **111**
    remedial cementing
    (*refer also to* cement job), **33**, **47**
centrifugal
    centrifugal pump or pumping, 176, **195**, **219**, **250**
    measurement in submerged centrifugal pumping, 211
CET (Cement Evaluation Tool), 46, **47**
CFE (Core Flow Efficiency), 62
charge
    shaped charge, **59**
check valve, 233, **239**, 290
checking
    checking , **33**, **163**
    checking tool, **269**
choke, 16, 118, **173**
    choke bean, 118
Christmas tree (*or* Xmas tree), 115, **118**, 170, **295**, **296**
circulating device, 142
clay stabilizer, 93
cleaning (*refer also to* clearing)
    cleaning the perforation, **74**
clearing, 34
    clearing fluid, **172**
    well clearing, **74**, **173**
cluster, 154, 195
coiled tubing *or* **coiled tubing unit,** 12, 95, 253, **276**, 305
collapse pressure, 126
collar
    die collar, 299
combination safety valve, 159
communication
    pay zone-borehole communication (*refer to* pay zone-borehole connection)
completion, 1, 11, 23, 32
    alternate selective completion, **30**
    cased hole completion, 25, **26**, **84**, **87**
    completion fluid, **40**, **163**
    completion report, 174
    conventional completion, 27

miniaturized completion, **31**
multiple zone completion, **28**, **31**
open hole completion, **25**, **82**
single zone completion, **27**, **31**
tubingless completion, 27, **30**
complexing (*or* sequestring) **agent**, 93
concentric valve
  mandrel with concentric valve, **237**
conductivity, 100
confirmation well, **3**, 12
coning, 8, 109
connection, **125**, **128**, **130**
  pay zone-borehole connection, **25**, **33**, **37**, **110**, **164**, **256**, **257**
  tubing-packer connection, **135**
consolidation, 77, **81**, **85**
continuous gas lift, 223, **225**, **231**, **241**, **251**
control
  control head, 265
  control line (*or* liner), 166, 168, 171, 240, 256
  sand control, **81**, **112**
conventional
  conventional completion (*refer also to* completion), **27**
  conventional mandrel, **236**
corrosion inhibitor, 93
coupling
  flow coupling, 117, 152
  quick-lock coupling, 266, 269
CPF (Closed Power Fluid), 200
cross, 118
  cross-linker, 93
crossover block, 201
curve
  IPR (Inflow Performance Response) curve, 21
  performance curve, **21**
  SIP (System Intake Performance) curve, 21
cutter
  explosive cutter, 303
  gage cutter, 269
  wireline cutter, 274
cuttings, 38

D

damage (*refer also to* skin)
  formation damage, 6, **38**, 91, **299**

deep treatment, 90
demulsifier, 93
depleted reservoir, 298
development well, **3**, **12**
diameter
  casing diameter, **43**
  drift diameter, **123**
  drilling diameter, 5, **43**
  EH (Entry Hole) diameter, 61
  ID (Inside Diameter), **123**
  nominal diameter, **122**, **128**, 146
  OD (Outside Diameter), 122, **123**, 182
die collar, 299
disconnection joint, 117, 153
diverting agent, 94, 99
dog leg, 305
downhole
  downhole accessories (*refer also to* downhole equipment), **281**, **290**
  downhole equipment (*refer also to* downhole accessories), **142**, **153**, **178**, **256**
  downhole tool, 274
drift diameter, 123
Drill Stem Test (*refer to* DST)
drillable packer (*refer to* permanent packer)
drilling, 4
  drilling and casing program, **5**
  drilling diameter, 5, **43**
  drilling fluid, **6**, **38**
  drilling in the pay zone, **6**, **37**
  drilling rig, **4**
drive mechanism, 7
DST (Drill Stem Test), 72
dummy valve, 143, 237
dump
  dump well, 7
  trash dump, 70
dynamic level, 186, 187
dynamometer, 211
dynamometric measurement, 209

E

echometer, 209
ECP (Effective Core Penetration), 62

INDEX

effective
    ECP (Effective Core Penetration), 62
    effective permeability (*refer also to* permeability), 78
EH (Entry Hole diameter), 61
Entry Hole diameter (*refer to* EH)
equalizer (*or* protector), **197**
equipment, 115
    downhole equipment (*refer also to* downhole accessories), **142, 153, 178, 256**
    equipment installation (*refer also to* equipment *and to* running procedure), **34**
EU (External Upset), 126
exploration well, 2, 12
explosive cutter, 303
extension
    mill-out extension, 138
External Upset (*refer to* EU)

F

fail safe, 120
FE (Flow Efficiency), 14
filter cake, 38
filtrate, 38
finder
    wireline finder, 275
fish, 254, 258
    wireline fish (*refer also to* fishing), 174
fishing (*refer also to* fish)**,** 255, **258**, 283
    fishing tool, **274**, **299**
fixed
    fixed casing pump, 201
    fixed insert pump, 201
    fixed pump, 201
flow
    flow capacity, **13**
    flow coupling, 117, 152
    FE (Flow Efficiency), 14
    gas flow, **15**
    oil flow, **13**
flowing well, 10, 115
fluid
    annular fluid (*or* packer fluid), **131**, **172**
    clearing fluid, **172**
    completion fluid, **40**, **163**

CPF (Closed Power Fluid), 200
drilling fluid, **6**, **38**
fluid-loss additive, 99
fracturing fluid, **97**
light fluid, 68
OPF (Open Power Fluid), 200
packer fluid (*or* annular fluid), **131**, **172**
treatment fluid, **94**
formation damage, 6, **38**, 91, **299**
formation plugging
    (*refer also to* plugging *and to* damage), 6, 14, 38, 90
frac
    acid frac, 97
fracture gradient, 96
fracturing
    fracturing fluid, **97**
    fracturing pressure (*or* breakdown pressure), 97
    hydraulic fracturing, **95**, 113
    measurement after fracturing, **104**
free
    free casing pump, 201
    free parallel pump, 201
    free pump, 201
friction reducer, 99
full bore (*refer to* landing nipple)

G

gage
    gage cutter, 269
    gage ring, 164
Gamma Ray (γR), 33, 73, 163, 165
gas lift, 7, 8, 19, 20, 43, 143, 159, 175, **222**, 254
    closed circuit gas lift, 224, 226
    continuous gas lift, 223, **225**, **231**, **241**, **251**
    gas-lift valve (*see below*)
    intermittent gas lift, 223, **241**
    measurement in gas lift, **247**
    open circuit gas lift, 224, 226
gas-lift valve, 233
    bellows valve, **233**, **234**
    casing operated bellows valve, 233
    production pressure (*or* tubing) operated valve, 234
    spring (*or* spring loaded) valve, **234**, **236**
    tubing (*or* production pressure) operated valve, 234
    unloading valve, **230**

gas
   gas anchor, **178**
   gas cap, 7, 20, 107
   gas lift (*see below*)
   gas lock, 178, **215**
   GLR (Gas Liquid Ratio), 240
   GOR (Gas Oil Ratio), 105, 197, 224

gelling agent, 99
GLR (Gas Liquid Ratio), 240
go-devil, 275
GOR (Gas Oil Ratio), 105, 197, 224
grab
   wireline grab, 275
grade, 122, **124**, 128, **130**, 183
gradient
   fracture gradient, 96
   pressure gradient, 227
gravel, 82, **87**
   gravel packing, **82**, **84**, **87**
gray valve, 295
groove
   locking groove, 146
gun (*refer also to* TCP: Tubing Conveyed Perforator), **65**
   casing gun, 66, 164
   through tubing gun, 68

H

hanger
   liner hanger, 5
Hardness Rockwell C (*refer to* HRC)
head
   casing head spool, 118, 161
   horsehead, 189
   injector head, 278, **279**
   pumping head, **186**
heading, 155
heavy
   heavy operation, **276**
   heavy pipe phase (*refer to* strip phase)
hesitation squeeze, 51, 59
high-pressure squeeze, 49
hold-down button, 139
hole
   cased hole completion, 25, **26**, 84, **87**
   open hole completion, 25, **82**

horizontal well, 107
   operation on horizontal wells, **303**
horsehead, 189
HRC (Hardness Rockwell C), 125
hydraulic
   hydraulic fracturing, **95**, 113
   hydraulic jar, 266, 267, 303
   hydraulic jet pump, **204**
   hydraulic packer, **139**, **169**, **170**
   hydraulic plunger pump, **199**
   hydraulic pumping, 176, **199**, **221**, **250**
   hydraulic unit, **193**

I

ID (Inside Diameter), **123**
impression block, 275
Improved Plow Steel (*refer to* IPS)
impulse factor, 189
Inflow Performance Response Curve (*refer to* IPR curve)
injection well, 3
injector head, 278, **279**
insert
   fixed insert pump, 201
inserted pump (*refer to* R pump)
Inside Diameter (*refer to* ID)
installation
   equipment installation (*refer also to* equipment *and to* running procedure), **34**
Instantaneous Shut in Pressure (*refer to* ISIP)
integral joint, 125, 289
intensifier
   jar intensifier, 303
intensity measurement, 211
interface between fluids, 8, **25**, **26**
intermittent gas lift, 223, **241**
intermitter, 241, 244
internal yield pressure, 126
IPR curve (Inflow Performance Response curve), 21
IPS (Improved Plow Steel), 262
ISIP (Instantaneous Shut In Pressure), 97

INDEX

J

J slot, 52, 142
JAM (Joint Analysed Make-up), 162
jar 267
    jar accelerator *or* jar intensifier, 303
    hydraulic jar, 266, 267, 303
    knuckle jar, 269
    mecanical jar, 267
    spang jar, 265, 267
    tubular jar, 267
jarring tool, 303
jet pumping, 204, **221**
jetting tool, 303
joint
    blast joint, 117, 153
    disconnection joint, 117, 153
    integral joint, 125, 289
    JAM (Joint Analysed Make-up), 102
    knuckle joint, 265, **269**
    premium joint, **126**, 130
    pup joint, 123, 167
    safety joint, 117, 153, 303
    slip joint, 117, 153
junk
    junk basket, 164
    junk catcher, 299

K

kick-off pressure, 231
kickover tool, 253, 266, 269
kill string, 10, 28
killed well (*refer also to* killing the well), 10, 254, **293**
    operation on killed wells, **293**
killing the well (*refer also to* killed well) **295**, **296**
knuckle
    jar, 269
    joint, 265, **269**

L

landing nipple, 117, **145**, **150**, **152**, 166, 291
    bottom no-go landing nipple, 146, **150**, 151, 152, 153
    full bore landing nipple, **146**, **148**, **149**, 151, 152
    no-go landing nipple (*refer to* bottom no-go landing nipple)
    ported landing nipple, **145**
    selective landing nipple, 146, **148**, 151, 152
    top no-go landing nipple, 146, **149**, 151, 152
latch, 233, 237
level
    dynamic level, 186, 187
    lost level, 299
    static level, 187
lift
    artificial lift, 7, 10, 19, **175**, **247**
    gas lift (*refer to* gas lift)
light
    light fluid, 68
    light operation, **258**
    light pipe phase (*refer to* snub phase)
liner, 5, 26, 43, 111
    (*or* control line), 166, 168, 171, 240, 256
    liner hanger, 5
    perforated liner, 25
live well, 258, 276
locator seal assembly, 135
lock, 146
    gas lock, 178, **215**
    lock mandrel, 145, **274**
    quick-lock coupling, 266, 269
locking groove, 146
log *or* **logging**
    cased hole logging, **163**
    production logging, **103**, 255
    (wireline) logging, 2, 33, 45, 73
long stroke unit, 286
loss, 299
lost level, 299
low-pressure squeeze, 50, **56**
lubricator, 259, **263**

M

macaroni string, 123, 289
magnet, 275
maintenance
    maintenance operation, **11**, **35**, **255**
    maintenance tool, **269**

mandrel
    conventional mandrel, **236**
    lock mandrel, 145, **274**
    mandrel with concentric valve, **237**
    side pocket mandrel, **150**, **236**
    valve mandrel, **236**
master valve (*refer also to* valve), 118, 255
matrix treatment, 90
measurement
    dynamometric measurement, **209**
    intensity measurement, **211**
    measurement after fracturing, **104**
    measurement in gas-lift, **247**
    measurement in submerged centrifugal pumping, **211**
    measurement on pumped wells, **207**
    measurement operation, **11**, **35**, **254**
    MWD (Measurement While Drilling), 303
mechanical jar, 267
mechanically set packer, 142
mechanism
    drive mechanism, 7
milling tool (*refer also to* tool), 303
mill-out extension, 138
miniaturized completion
    (*refer also to* completion), **31**
mobility, 24, 25, 26
Moyno pump or **pumping,** 176, **206**
mud acid, 93
multiple zone completion
    (*refer also to* completion), **28, 31**
MWD (Measurement While Drilling), 303

N

NACE (National Association of Corrosive Engineers), 124
National Association of Corrosive Engineers (*refer to* NACE),
nest, 274
nipple (*refer to* landing nipple),
no-go
    bottom no-go (*refer to* landing nipple),
    no-go ring, 149
nominal
    nominal diameter, **122**, **128**, 146

    nominal weight, **123**, 128, **130**
Non Upset (*refer to* NU)
NU (Non Upset), 126

O

O ring, 263
OAP (Over All Penetration), 62
observation well, 3
OD (Outside Diameter), 122, **123**, 182
offshore, 4, 10, 30, 120, 154, 159, 240, 289
open hole completion
    (*refer also to* completion), **25, 82**
Open Power Fluid (*refer to* OPF)
operation
    heavy operation, **276**
    light operation, **258**
    maintenance operation, **11**, **35**, **255**
    measurement operation, **11**, **35**, **254**
    operations on horizontal wells, **303**
    operations on killed wells, **293**
    operations on subsea wells, **305**
    workover operation, **11**, **35**, **256**
OPF (Open Power Fluid), 200
Outside Diameter (*refer to* OD)
Over All Penetration (*refer to* OAP)
Over Power Fluid (*refer to* OPF)
overbalanced pressure perforating, 66
overflush, 93
overshot, 275, 299
overtravel, 185, 219

P

packer, 10, 27, 28, 29, 30, 43, 117, **131**, 175, 254
    drillable packer (*refer to* permanent packer)
    hydraulic packer, **139**, **169**, **170**
    mechanically set packer, **142**
    packer fluid (*or* annular fluid), **131**, **172**
    permanent packer, **132**, **133**, **138**, **164**, **167**
    retrievable packer, **132**, **138**
    squeeze packer, 51, 53
packing, 146, 237
pad, 102
paralell
    free paralell pump, 201

pay zone
    drilling in the pay zone, **6**, **37**
    pay zone-borehole connection, **25**, **33**, **37**, **110**, **164**, **256**, **257**
penetration
    TCP (Total Core Penetration), 61
    TTP (Total Target Penetration), 62
perforated
    perforated liner, 25
    perforated tube, 153
perforating, 59, 111, 164
    overbalanced pressure perforating, **66**
    perforating after equipment installation, **68**
    perforating before equipment, **66**
    perforating method, **65**
    TCP (Tubing Conveyed Perforator) perforating, **70**
    underbalanced pressure perforating, **68**
perforation, 79
    cleaning the perforation, **74**
performance curve, 21
permanent packer, 132, **133**, **138**, **164**, **167**
permeabillity, 13, 20, **90**, **107**
    absolute permeability, **78**
    effective permeability, **78**
    relative permeability, **78**
phase
    snub phase or snubbing phase or lightpipe phase, 286, 292, 293
    strip phase or stripping phase or heavy pipe phase, 286
PI (*refer also to* productivity), 13, 14, **20**, 38, 43, 107, 186, 226
pipe
    heavy pipe phase (*refer to* strip phase)
    light pipe phase (*refer to* snub phase)
    pipe ram (*refer also to* ram), 281
plug, 294
    bridge plug, 51, **56**
plugging (*refer also to* damage), 66, 68, 94, 297
    formation plugging, 6, 14, 38, 90
plunger, 182
    plunger pumping, **199**, **221**
    plunger valve, 176
polished rod, 185, 218
pony rod, 184
poppet valve, 154

port, 233
    port collar, 113
ported landing nipple (*refer also to* landing nipple), **145**
positioning tool (*refer to* kickover tool)
preflush, 92
pre-job meeting, 103
premium joint, 126, 130
pressure
    bottomhole pressure (P_{BH}), 13, **17**
    breakdown or fracturing pressure, 97
    burst or internal yield pressure, 126
    collapse pressure, 126
    fracturing or breakdown pressure, 97
    high-pressure squeeze, **49**
    internal yield or burst pressure, 126
    ISIP (Instantaneous Shut in Pressure), 97
    kick-off pressure, 231
    low-pressure squeeze, **50**, **56**
    pressure differential safety valve, **154**
    pressure gradient, 227
    pressure operated safety valve, **155**
    propagation pressure, 97
    reservoir pressure (P_R), **6**, 13, **20**
    WP (Working Pressure), 120
preventer (*refer to* BOP)
production, 10
    production casing, **6**
    production logging, **103**, 255
    production pressure (*or* tubing) operated valve (*refer also to* gas lift), 234
    production string (*refer to* tubing)
    production well, **3**
    production wellhead, 115, **117**, **171**, **254**, **255**, **256**
    selective production (*refer also to* selectivity), 23
productivity, 63, **78**, **90**
    PI (Productivity Index), 13, 14, **20**, 38, 43, 107, 186, 226
program
    drilling and casing program, **5**
propagation pressure, 97
propping agent, 97, **99**
proprietary grade, 124

protector (*or* equalizer), **197**
PU (Pumping Unit), 188
pulling
  pulling tool, **272**
  pulling unit (*refer to* servicing unit)
pump (*refer also to* pumping)
  bottomhole pump, **180**
  centrifugal pump, **195**
  fixed casing pump, 201
  fixed insert pump, 201
  free casing pump, 201
  free parallel pump, 201
  hydraulic jet pump, **204**
  hydraulic plunger pump, **199**
  inserted pump (*refer to* R pump)
  Moyno pump or pumping, **206**
  pump down stinger, 304
  R (Rod) pump, 180, 182
  T (Tubing) pump, 182
  turbine pump, **205**
pumped well (*refer also to* pumping)
  measurement on pumped wells, **207**
pumping (*refer also to* pump *and to* pumped well), 7, 19, **175, 276**
  centrifugal pumping, 176, **195, 219, 250**
  defining a pumping installation, **215**
  hydraulic pumping, 176, **199, 221, 250**
  jet pumping, **204, 221**
  Moyno pumping, 176
  plunger pumping, **199, 221**
  pumping head, **186**
  pumping parameter, **186**
  PU (Pumping Unit), **188**
  rod pumping, **176, 215, 249**
  submerged centrifugal pumping (*refer to* centrifugal pumping)
  sucker rod pumping (*refer to* rod pumping)
pup joint, 123, 167
PVT (Pressure – Volume – Temperature), **81**

Q

quick-lock coupling, 266, 269

R

R pump (Rod pump), 180, 182

ram
  blind ram, 290
  pipe ram, 281
  shear ram, 281
  slip ram, 281
range, 123
reaming out, 84, 87
Recommended Practive (*refer to* RP)
relative permeability
    (*refer also to* permeability), **78**
release stud, 135, 165
releasing spear, 303
remedial cementing (*refer also to* cementing *and to* cement job), 33, **47**
Remote Operated Vehicle (*refer to* ROV)
reservoir, 1, **6**
  depleted reservoir, **298**
  reservoir pressure P_R, **6**, 13, **20**
retrievable
  retrievable packer, 132, **138**
  TR (Tubing Retrievable) safety valve, **157,** 256
  WLR (WireLine Retrievable) safety valve, **157,** 161, 255, 256
reversing tool, 303
ring
  gage ring, 164
  no-go ring, 149
  O ring, 263
rockwell, 125
rod
  polished rod, 185, 218
  pony rod, 184
  R (rod) pump, 180, 182
  rod pumping, **176, 215, 249**
  sucker rod, **182,** 266
  sucker rod string, 184, **218**
rope socket, 265, 266
ROV (Remote Operated Vehicle), 306
RP (Recommended Practice), **60,** 217
running
  running procedure, **161**
  running tool, 147, **272**

S

S (*refer to* Skin)

INDEX

safety (*refer also to* BOP), **10**, **37**, **38**, **72**, **280**, **289**, **294**
  safety barrier, 170
  safety BOP, 290
  safety joint, 117, 153, 303
  safety valve (*see below*)

safety valve
  ambient safety valve, 155
  combination safety valve, **159**
  pressure differential safety valve, **154**
  pressure operated safety valve, **155**
  SCSSV (Surface Controlled Subsurface Safety Valve), **156**, **166**, **169**, **172**
  SSCSV (SubSurface Controlled subsurface Safety Valve), **154**
  SSSV (SubSurface Safety Valve), 117, **153**, 255, **256**
  SSTA (SubSurface Tubing Annulus) safety valve, 157, **159**, 239
  SSV (Surface Safety Valve), 120, 255
  TM (Tubing Mounted) safety valve (*refer to* TR)
  TR (Tubing Retrievable) safety valve, **157**, 256
  velocity safety valve, 154
  WLR (WireLine Retrievable) safety valve, **157**, 161, 255, 256

sand
  sand bailer, 272
  sand control, **81**, **112**

sand-out (*or* screen-out), 102, 105

scraper, 33, 59, 163, 296

scratcher, 269

screen, 82, **87**

screen-out (*or* sand-out), 102, 105

SCSSV (Surface Controlled Subsurface Safety Valve), **156**, **166**, **169**, **172**

seal assembly
  anchor seal assembly, **135**
  locator seal assembly, **135**

seal bore, 146

section mill, 303

selective
  selective completion (*refer to* alternate selective completion)
  selective landing nipple (*refer to* landing nipple)
  selective production, 23

selectivity, 26
separation sleeve, 166, 172
sequestring (*or* complexing) **agent,** 93
servicing
  servicing unit (*refer also to* workover unit), 249, 293, **295**, **296**
  well servicing (*refer also to* workover), **253**
shaped charge, 59
shear ram (*refer also to* ram), 281
short stroke unit, 286
shot per foot (*refer to* SPF)
side pocket mandrel, 143, **236**
sidetracking, 257
Simphor (Système d'Information et de Mesure en Puits HORizontaux),
single zone completion (*refer also to* completion), **27**, **31**
sinker bar, 266
SIP curve (System Intake Performance curve)**,** 21
skin (*refer also to* damage)
  skin effect, 14, 77
  skin factor (S), 14
slack off, 170
sleeve,
  separation sleeve, 166, 172
Sliding Side Door (*refer to* SSD)
Sliding Sleeve (*refer to* SS)
slip
  slip joint, 117, 153
  slip ram (*refer also to* ram), 281
sludge, 38
snub phase, 286, 292, 293
snubbing or **snubbing unit,** 12, 95, 253, **283**
  snubbing phase (*refer to* snub phase)
socket
  rope socket, 265, **266**
sonolog, 209
spacing out, 166
spang jar, 265, 267
SPEC (Specification), 180, 262
SPF (Shot Per Foot), 63
SPM
  Side Pocket Mandrel, **143**, **236**
  Stroke Per Minute,

spool
    casing head spool, 118, 161
spring (*or* sping loaded) **valve**
    (*refer also to* gas lift), 234, 236
squeeze, 49, **51**, **58**, **295**, **297**
    hesitation squeeze, 51, 59
    high-pressure squeeze, **49**
    low-pressure squeeze, **50**, **56**
    squeeze packer, 51, 53
SRPR (Sucker Rod Pumping Research), 217
SS (Sliding Sleeve) *or* **SSD** (Sliding Side Door), **117**, 143
SSCSV (SubSurface Controlled subsuface Safety Valve), **154**
SSD (Sliding Side Door) *or* **SS** (Sliding Sleeve), **117**, **143**
SSSV (SubSurface Safety Valve), 117, **153**, 255, **256**
SSTA (Subsurface Tubing-Annulus) Safety valve, 157, **159**, 239
SSV (Surface Safety Valve), 120, 255
stabbing guide, 165
standing valve, 168, 176
static level, 187
stem, 265
    wireline stem, 265, **266**
stimulation, 20, **88**, **113**
stinger, 54
    pump down stinger, 304
storm choke, 154
strength
    body yield strength, 124
stress
    body yield stress, 124
string
    kill string, 10, 28
    macaroni string, 123, 289
    production string (*refer to* tubing)
    sucker rod string, 184, **218**
strip phase, 286
stripper, 280, 289, 292
stripping
    stripping BOP, 289, 292
    stripping phase (*refer to* strip phase)
stroke
    long stroke unit, 286
    short stroke unit, 286
stuffing box, 263
submerged centrifugal pump *or* **pumping**
    (*refer to* centrifugal pump or pumping)
submergence, 187
subsea well
    operation on subsea wells, **305**
SubSurface Controlled subsurface Safety Valve (*refer to* SSCSV)
SubSurface Safety Valve (*refer to* SSSV)
SubSurface Tubing-Annulus safety valve
    (*refer to* SSTA safety valve)
sucker
    sucker rod, **182**, 266
    sucker rod pumping (*refer to* rod pumping)
    sucker rod string, 184, **218**
Surface Controlled Subsurface Safety Valve (*refer to* SCSSV)
Surface Safety Valve (*refer to* SSV)
surfactant, 93, 99
swab valve, 118
swage, 168
swaging tool, 269
System Intake Performance curve
    (*refer to* SIP curve)

T

T pump (Tubing pump), 182
tail pipe, 51, 53
taper tap, 303
TCP,
    TCP (Tubing Conveyed Perforator) perforating, **70**
    Total Core Penetration, 61
    Tubing Conveyed Perforator (*refer also to* gun), 65, **70**, 85
TDH (Total Dynamic Head), 198
temporary plugging agent, 94, 299
tensiometer, 262
test port, 171
testing
    well testing, 2, 3, **34**, **81**

INDEX

TFL (Through Flow Line), 306
through
    Through Flow Line (*refer to* TFL)
    through tubing gun, 68
thru tubing gun (*refer to* through tubing gun)
TM (Tubing Mounted) **safety valve** (*refer to* TR)
tool
    catch tool, 299, 303
    checking tool, **269**
    downhole tool, **274**
    fishing tool, **274**, **299**
    jarring tool, 303
    jetting tool, 303
    kickover tool, **253**, 266, 269
    maintenance tool, **269**
    milling tool, 303
    pulling tool, **272**
    reversing tool, 303
    running tool, 147, 272
    tool trap, 259, **264**
    washover tool, 303
    wireline tool, **269**
top no-go (*refer to* landing nipple)
torque turn, 166
Total Core Penetration (*refer to* TCP)
Total Target Penetration (*refer to* TTP)
TR (Tubing Retrievable) **safety valve, 157**, 256
transit time, 47
trash dump, 70
travelling valve (*or* plunger valve), 176
treating (*refer also to* treatment), **34**, **37**
treatment (*refer also to* treating)
    deep treatment, **90**
    matrix treatment, **90**
    other treatment, **90**
    treatment fluid, **94**
tree
    Christmas tree (*or* Xmas tree), 115, **118**, 170, **295**, **296**
    tree cap, 116
TTP (Total Target Penetration), 62
tubing (*or* production string) 10, **27**, 43, **113**, 117, **122**, **128**, 165, **168**, **169**, 175, 254, **255**, 256
    TCP (Tubing Conveyed Perforator), 65, **70**, 85
    TM (Tubing Mounted: *refer to* TR)

T (Tubing) pump, 182
through tubing gun (*or* thru tubing gun), 68
tubing anchor, 159, **178**
tubing hanger, 118, **168**, 170
tubing (*or* production pressure) operated valve (*refer also to* gas lift), 234
TR (Tubing Retrievable) safety valve, **157**, 256
tubing-head or **tubing-head spool,** 117, 118, 161, **168**, 171, 201, **252**, 312
tubingless completion (*refer also to* completion), 27, **30**
tubing-packer connection, 135
tubular jar, 267
turbine pump, 205
TWCV (Two Way Check Valve), 172
Two Way Check Valve (*refer to* TWCV)

U

underbalanced pressure perforating, 68
unit
    B unit, 190
    C unit, 190
    coiled tubing unit (*or* coiled tubing), 12, 95, 253, **276**, 305
    hydraulic unit, **193**
    long stroke unit, 286
    PU (pumping unit), **188**
    servicing unit, 249, 293, **295**, **296**
    short stroke unit, 286
    snubbing unit (*or* snubbing), 12, 95, 253, **283**
    walking beam unit, **188**
    workover unit (*or* rig), 293, **295**, **296**
unloading valve (*refer also to* gas lift), **230**
upset
    EU (External Upset), 126
    NU (Non Upset), 126

V

valve
    back pressure valve, 118, 171, 294
    check valve, 233, **239**, 290
    dummy valve, 143, 237
    gas lift valve (*refer to* gas lift)
    mandrel with concentric valve, **237**
    master valve, 118, 255
    plunger valve (*or* travelling valve), 176

poppet valve, 154
safety valve (*refer to* safety)
standing valve, 168, 176
swab valve, 118
travelling valve (*or* plunger valve), 176
valve mandrel, **236**
wing valve, 118

VAM, 126

velocity safety valve, 154

viscosity, 13, 21, **79**

W

walking beam unit, 188, 190

washing
acid washing, **75**
washing tool, **75**, 85

washover tool (*refer also to* tool), 303

water block, 91

Water Oil Ratio (*refer to* WOR)

wave equation method, 218

wear bushing, 165

weight
nominal weight, **123**, 128, **130**
weight indicator, **262**

well
appraisal well, **3**, 12
confirmation well, **3**, 12
development well, **3**, 12
dump well, 7
exploration well, **2**, 12
flowing well, **10**
horizontal well (*refer also to* operation on horizontal well), **107**
gas well, **19**
injection well, 3
killed well (*refer also to* killing the well), 10, 254, **293**
killing the well (*refer also to* killed well), **295**, **296**
live well, 258, 276
observation well, 3
oil well, **17**
operation on horizontal wells (*refer also to* horizontal well), **303**
operation on killed wells, **293**
operation on subsea wells, **305**
production well, 3

well clearing, **74**, **173**
well servicing (*refer also to* workover), **253**
well stimulation, 20, **88**, **113**
well testing, 2, 3, **34**, **81**

wellhead or **production wellhead,** 115, **117**, 161, **171**, **254**, **255**, **256**

wettability, 78

wetting agent, 93

window, 28, 86

wing valve, 118

wire, 265

wireline *or* **wireline unit,** 12, 142, 253, **258**
wireline cable, 262
wireline cutter, 274
wireline finder, 275
wireline fish (*refer also to* fishing), 174
wireline grab, 275
wireline logging (*refer also to* logging), 2, 33, 45, 73
wireline stem, 265, **266**
wireline string, **265**
wireline tool, **269**

WLR (WireLine Retrievable) **safety valve, 157**, 161, 255, 256

WOR (Water Oil Ratio), 197, 226

working pressure (*refer to* WP)

WP (Working Pressure), 120

workover (*refer also to* well servicing), 1, 194, **253**
workover operation, **11**, **35**, **256**

workover rig or unit (*refer also to* servicing unit), **293**, **295**, **296**

X

Xmas tree (*or* Christmas tree), 115, **118**, **170**, **295**, **296**

Y

yield
body yield strength, 126
body yield stress, 124
internal yield pressure, 126

ACHEVE D'IMPRIMER
EN JANVIER 2011
PAR L'IMPRIMERIE JOUVE - 53100 MAYENNE
N° d'impression : 598778S
dépôt légal : Octobre 2004
IMPRIME EN FRANCE